DESCRIBING SOCIOECONOMIC FUTURES
for Climate Change Research and Assessment

REPORT OF A WORKSHOP

Panel on Socioeconomic Scenarios for
Climate Change Research and Assessment

Committee on the Human Dimensions of Global Change

Division of Behavioral and Social Sciences and Education

NATIONAL RESEARCH COUNCIL
OF THE NATIONAL ACADEMIES

THE NATIONAL ACADEMIES PRESS
Washington, D.C.
www.nap.edu

THE NATIONAL ACADEMIES PRESS   500 Fifth Street, N.W.   Washington, DC 20001

NOTICE: The project that is the subject of this report was approved by the Governing Board of the National Research Council, whose members are drawn from the councils of the National Academy of Sciences, the National Academy of Engineering, and the Institute of Medicine. The members of the committee responsible for the report were chosen for their special competences and with regard for appropriate balance.

This project was supported by the National Science Foundation through award number SES-1003678, with contributions from the U.S. Department of Energy and the National Oceanic and Atmospheric Administration. The National Institute for Environmental Studies (Japan) provided travel support for several participants. Support of the work of the Committee on the Human Dimensions of Global Change is provided by a consortium of federal agencies through a contract from the National Aeronautics and Space Administration (Number NNH07CC79B) and by a grant from the U.S. National Science Foundation (Number BCS-0436369). Any opinions, findings, conclusions, or recommendations expressed in this publication are those of the author(s) and do not necessarily reflect the views of the sponsors.

International Standard Book Number-13:   978-0-309-16144-2
International Standard Book Number-10:   0-309-16144-4

Additional copies of this report are available from the National Academies Press, 500 Fifth Street, N.W., Lockbox 285, Washington, DC 20055; (800) 624-6242 or (202) 334-3313 (in the Washington metropolitan area); Internet http://www.nap.edu.

THE NATIONAL ACADEMIES
Advisers to the Nation on Science, Engineering, and Medicine

The **National Academy of Sciences** is a private, nonprofit, self-perpetuating society of distinguished scholars engaged in scientific and engineering research, dedicated to the furtherance of science and technology and to their use for the general welfare. Upon the authority of the charter granted to it by the Congress in 1863, the Academy has a mandate that requires it to advise the federal government on scientific and technical matters. Dr. Ralph J. Cicerone is president of the National Academy of Sciences.

The **National Academy of Engineering** was established in 1964, under the charter of the National Academy of Sciences, as a parallel organization of outstanding engineers. It is autonomous in its administration and in the selection of its members, sharing with the National Academy of Sciences the responsibility for advising the federal government. The National Academy of Engineering also sponsors engineering programs aimed at meeting national needs, encourages education and research, and recognizes the superior achievements of engineers. Dr. Charles M. Vest is president of the National Academy of Engineering.

The **Institute of Medicine** was established in 1970 by the National Academy of Sciences to secure the services of eminent members of appropriate professions in the examination of policy matters pertaining to the health of the public. The Institute acts under the responsibility given to the National Academy of Sciences by its congressional charter to be an adviser to the federal government and, upon its own initiative, to identify issues of medical care, research, and education. Dr. Harvey V. Fineberg is president of the Institute of Medicine.

The **National Research Council** was organized by the National Academy of Sciences in 1916 to associate the broad community of science and technology with the Academy's purposes of furthering knowledge and advising the federal government. Functioning in accordance with general policies determined by the Academy, the Council has become the principal operating agency of both the National Academy of Sciences and the National Academy of Engineering in providing services to the government, the public, and the scientific and engineering communities. The Council is administered jointly by both Academies and the Institute of Medicine. Dr. Ralph J. Cicerone and Dr. Charles M. Vest are chair and vice chair, respectively, of the National Research Council.

www.national-academies.org

Preface

The implications of climate change for the environment and society depend not only on the rate and magnitude of climate change, but also on changes in technology, economics, lifestyles, and policy that will affect the capacity both for limiting and adapting to climate change. The workshop that is the subject of this report was organized by the National Research Council's (NRC's) Committee on the Human Dimensions of Global Change and the Climate Research Committee to initiate a dialogue among interested researchers to explore the requirements for descriptions of socioeconomic and environmental futures to complement climate scenarios. Participants came from several countries and considered approaches and methodological issues in developing socioeconomic scenarios, the forces and uncertainties that will affect adaptation potential and vulnerability as well as emissions and mitigation potential, and the possible elements of a research plan to advance development of socioeconomic scenarios and narratives.

The objectives of the workshop were to review the state of science for considering socioeconomic changes over long time frames; clarify definitions and concepts to facilitate communication across research communities; brainstorm about driving forces and key uncertainties that will affect impacts, adaptation, and vulnerability and mitigation in the future; and consider research needs and the elements of a strategy for describing socioeconomic and environmental futures for climate change research and assessment. Specifically, the participants reviewed narrative and quantitative methods from a range of disciplines for developing long-term scenarios of socioeconomic futures; identified key factors that might influence adaptation, mitigation, and the environment in the coming decades and that need to be covered in future scenarios; discussed a new process for scenario development that uses representative concentration pathways (RCPs) of future forcing and examined the range of socioeconomic assumptions in model runs consistent with the RCPs; and shared prior experience in the use of narratives and scenarios.

The workshop addressed a number of specific methodological challenges and opportunities. First, any assessment of options to prepare for a changing climate requires not only current data on socioeconomic, climate, and other natural conditions, but also projections that extend for decades (centuries for some types of effects, such as sea level rise). Projections on these time scales challenge conventional scientific methods, and thus it is important to develop and apply socioeconomic scenarios consistent with their proper uses and limits, including a clear understanding that scenarios are not predictions but rather sets of internally consistent assumptions for testing the robustness of potential strategies to a range of potential futures.

Second, for assessments that seek to compare and synthesize information across different locations or systems, there is an additional need to provide an internally consistent set of data for diverse scenario elements—socioeconomic conditions, emissions, climate, broader environmental circumstances, and resources for responses. In previous assessments, both in the United States and internationally (e.g., the assessments of the Intergovernmental Panel on Climate Change [IPCC] or the Millennium Ecosystem Assessment), developing, disseminating, and applying consistent baselines and scenarios across scales from global to local have posed substantial challenges.

Another challenge in developing scenarios for an assessment is doing so in a way that blends scientific knowledge and input from users so that the scenarios are relevant to their concerns and will illuminate the consequences of different choices under their control in the context of broader uncertainties. Scenarios need to provide just the right amount of guidance and core information to facilitate coordination without overspecifying conditions or providing information that is irrelevant, lacks support from key stakeholders, or is not embedded in relevant institutional context. Again, past assessments have not been as successful as they might be on this score.

Over time, a variety of techniques to develop scenarios have been used, including temporal and spatial analogues and model-based scenarios. Traditional modeling approaches start from initial conditions and project forward, whereas other approaches identify desired future conditions and develop pathways for arriving at them. There have been advances in the methods available for providing climate information at finer scales of resolution (e.g., statistical and dynamical downscaling methods), but less attention has been given to preparing quantitative and narrative socioeconomic information. Advances in computing capacity are making development of probabilistic scenarios a reality. Recently, the research community developed a new "parallel approach" for developing integrated sets of socioeconomic, climate, and environmental scenarios, which has at least two potentially useful attributes: (1) the introduction of climate scenarios focused on approximately the next three decades, and (2) more flexibility to create socioeconomic scenarios that are embedded in consistent global narratives but that focus on the needs of specific decision makers and locations.

These new techniques and developments provide many options, but it remains to be seen how they can best be used, given inherent challenges. Central motives for holding this workshop were to explore the current state of science in scenario development and application and to discuss opportunities for a next round of assessments, including those of the IPCC and the U.S. Global Change Research Program. The workshop succeeded in raising and exploring these issues and in suggesting new lines of research needed to prepare for development of new socioeconomic scenarios to support future integrated assessments. Consistent with its charge, the panel did not attempt to come to consensus on recommendations or a specific research agenda.

Participants in the workshop identified a number of research needs and opportunities that are described in the report. One particularly important issue is additional research on socioeconomic scenarios for local and regional vulnerability assessments with different degrees of coupling to the global context of the RCPs. Developing such geographically "nested" scenarios will require a better understanding of the nature of interdependence between global trends and local adaptation and mitigation potential. Institutionally, additional coordination and information exchange, integration of data systems, and support for users are needed to realize the potential for increased collaboration that the new RCP scenario process presents. A wider range of insights will be developed if researchers and users from developing countries are integrated into the process to explore interactions among development strategies, adaptation, and mitigation.

The workshop was intended not only to identify research needs and opportunities, but also to support the process of planning the next and fifth assessment report of the Intergovernmental Panel on Climate Change. I wish to thank several leaders of IPCC, Kris Ebi, Ottmar Edenhofer, Chris Field, and Patrick Matschoss, and additional IPCC participants for their engagement. I thank the members of the Panel on Socioeconomic Scenarios for Climate Change Assessments, Kris Ebi, Kathy Hibbard, Anthony Janetos, Mikiko Kainuma, Ritu Mathur, Nebojsa Nakićenović, and Thomas Wilbanks, who developed the structure for the workshop and selected the participants. Presenters and participants endured "snowmaggedon" in Washington during early February 2010 and contributed their insights and knowledge to a lively and productive discussion. Finally, special thanks are due to Paul Stern, director of the Committee on Human Dimensions of Global Change, and Linda DePugh, of NRC, for their tireless efforts to organize the workshop.

This report has been reviewed in draft form by individuals chosen for their diverse perspectives and technical expertise, in accordance with procedures approved by the Report Review Committee of the NRC. The purpose of this independent review is to provide candid and critical comments that will assist the institution in making its published report as sound as possible and to ensure that the report meets institutional standards for objectivity, evidence, and responsiveness to the study charge. The review comments and draft manuscript remain confidential to protect the integrity of the deliberative process. We wish to thank the following individuals for their review of this report: Karen Fisher-Vanden, Department of Agricultural Economics and Rural Society, Pennsylvania State University; Tom Kram, Global Sustainability and Climate Unit, Netherlands Environmental Assessment Agency; and Brian C. O'Neill, Climate and Global Dynamics Division and Integrated Science Program, National Center for Atmospheric Research (NCAR), Boulder, CO.

Although the reviewers listed above have provided many constructive comments and suggestions, they were not asked to endorse the conclusions or recommendations nor did they see the final draft of the report before its release. The review of this report was overseen by Edward Parson, School of Law, University of Michigan. Appointed by the NRC, he was responsible for making certain that an independent examination of this report was carried out in accordance with institutional procedures and that all review comments were carefully considered. Responsibility for the final content of this report rests entirely with the authoring committee and the institution. Nonetheless, we thank the reviewers and the review coordinator for their diligent analysis and scrupulous comments, which have significantly improved the quality of the report.

Richard H. Moss, *Chair*
Panel on Socioeconomic Scenarios for
Climate Change Research and Assessment

Contents

1

Introduction

The Workshop on Describing Socioeconomic Futures for Climate Change Research and Assessment was organized in response to increasing recognition by the international research community working to analyze the consequences of climate change that improved socioeconomic scenarios are needed to understand climate change vulnerabilities and adaptive capacity. The need for improved analysis of feedbacks between human and climate systems was one of the themes that emerged from an international workshop organized by the National Research Council's (NRC's) Committee on the Human Dimensions of Global Change to consider lessons learned about analysis of climate change vulnerability, impacts, and adaptation from the experience of Working Group II in the Fourth Intergovernmental Panel on Climate Change (IPCC) assessment (National Research Council, 2009b). The need is pressing, both in relation to the IPCC's future tasks and to the research communities working on projecting and considering the long-term impacts of climate change.

The workshop was structured to combine invited presentations and discussions among the participants. The workshop, held on February 4-5, 2010, drew people from a wide variety of disciplines and international perspectives. The workshop agenda and a list of participants appear in Appendix A, and biographical sketches of panel members and staff appear in Appendix B.

PLAN OF THE REPORT

This report is a summary of the presentations at the workshop and the discussions flowing from the presentations during the sessions outlined in the agenda. It is important to be specific about its nature: the report documents the information presented in the workshop presentations and discussions. The report is confined to the material presented by the workshop speakers and participants. Neither the workshop nor this summary is intended as a comprehensive review of what is known about the topic, although it is a general reflection of the literature. The presentations and discussions were limited by the time available for the workshop.

Although this report was prepared by the panel, it does not represent findings or recommendations that can be attributed to the panel members. The report summarizes views expressed by workshop participants, and the panel is responsible only for its overall quality and accuracy as a record of what transpired at the workshop. Also, the workshop was not designed to generate consensus conclusions or recommendations but focused instead on the identification of ideas, themes, and considerations that contribute to understanding the topic.

INTRODUCTORY COMMENTS

Thomas Wilbanks, chair of NRC's Committee on the Human Dimensions of Global Change, welcomed the participants on behalf of the committee and the Climate Research Committee, which jointly planned and organized the workshop. He pointed out the importance of getting the science right for the next assessment of the Intergovernmental Panel on Climate Change. He noted that publication of the first peer-reviewed publication on representative concentration pathways (RCPs) was scheduled for February 11 in *Nature* (Moss et al., 2010).[1]

---

[1]RCPs involve a new approach to scenario development that recognizes that many scenarios of socioeconomic and technological development can lead to the same pathways of radiative forcing (changes in the balance of incoming and outgoing radiation to the atmosphere caused by changes in the concentrations of atmospheric constituents). Selecting a few RCPs for emphasis allows researchers to develop scenarios for the different ways the world might reach those RCPs and to consider the consequences of climate change when those RCPs are reached via specific scenarios. This approach has been proposed to increase research coordination and reduce the time needed to generate useful scenarios.

WORKSHOP OBJECTIVES, CONCEPTS, AND DEFINITIONS

Richard H. Moss

Richard Moss, chair of the Panel on Socioeconomic Scenarios for Climate Change Impact and Response Assessments, the organizing panel, described the workshop agenda and objectives. He emphasized that what will matter in the future is not only how much climate changes, but also how socioeconomic futures develop. As an example, he described a World Wildlife Fund project in Amazonia, intended to identify refugia[2] for protected species. He said the project had to consider changes in both climate and local socioeconomic drivers (e.g., changes in settlements, infrastructure, livestock production) and the interactions of all these factors. Socioeconomic scenarios were not readily available. He noted that important cross-scale effects need to be taken into account, citing the example of how wildlife habitats are affected by global markets, national policy, local changes, and changes in habitats and livelihoods.

Moss said that the workshop would examine how well scenarios used in climate change research reflect fundamental understanding of socioeconomic processes and change. People need to distinguish what is known from what is unknown and what is unknowable and also to characterize the level of confidence. He noted that there are many tools for analysis under uncertainty, one of which is scenarios.

Moss defined scenarios as plausible descriptions of how specific aspects of the future might unfold. Climate research uses many kinds of scenarios (socioeconomic, emissions, climate, environmental, vulnerability, etc.). He emphasized that scenarios are not predictions. He noted that synthesis requires coordination and that scenarios have a big role to play in coordinating different kinds of analysis.

Moss identified four objectives for the workshop:

1. assessing the state of the art/science in describing possible futures (using the best social science knowledge in ways that meet the needs of stakeholders);
2. supporting the IPCC and other assessments by advancing the framework for creating new scenarios and by identifying research needs and next steps;
3. thinking about the "drivers" of both emissions/mitigation and vulnerability/adaptation, including, in the case of vulnerability, the drivers of exposure, sensitivity, and adaptive capacity; and
4. promoting dialogue across research communities.

---

[2]Refugia are areas where special environmental circumstances have enabled a species or a community of species to survive after extinction in surrounding areas.

Moss noted the need for common definitions of certain terms but acknowledged that the research community has not yet coalesced around a single vocabulary for this area. For example, he noted that the terms "narrative" and "story line" both refer to detailed descriptions of the sequence of events that provide the logic for a quantitative scenario.

ADVANCING THE STATE OF SCIENCE FOR PROJECTING SOCIOECONOMIC FUTURES

Thomas J. Wilbanks

Thomas Wilbanks spoke about the importance of good descriptions of socioeconomic futures. Such descriptions are needed to integrate with climate projections on the same time scales. For example, the RCP report (Moss et al., 2008) called for a library of socioeconomic scenarios to go with climate scenarios. In the IPCC Fourth Assessment process, developing countries made strong calls for socioeconomic scenarios. Integrated assessment models (IAMs) project greenhouse gas emissions, which earth system models use as inputs to their climate projections.[3] These in turn are inputs to impact, adaptation, and vulnerability (IAV) analyses, which in turn feed back into emissions. Thus, the scientific communities that do IAM and IAV both have strong interest in improving the scientific base for descriptions of the socioeconomic future.

The scientific basis for the scenarios generated for the IPCC Special Report on Emissions Scenarios and for IAMs was developed from early work at the International Institute for Applied Systems Analysis and the work at the National Academy of Sciences for the report *Our Common Journey* (National Academy of Sciences, 1999). But many scientists question the basis of this work. For example, the core social sciences are generally skeptical of efforts to make socioeconomic projections far into the future. Wilbanks discussed an effort, in which he participated, to estimate coastal populations at risk from climate change in 2080. Such populations will depend on demographic and economic changes, as well as risk management responses in the interim. He said that the science and art of long-term socioeconomic projections are not equivalent to those of climate scenarios. Some of the reasons are that very little investment has been made in such work, that there are so many variables to analyze, and that there are no professional rewards for social scientists who try to do this kind of work. Consequently, the estimates used are based on very simple assumptions. Wilbanks said that projections are fairly commonly

---

[3]IAMs integrate socioeconomic and physical aspects of climate change, typically for the purpose of assessing policy options.

made as far into the future as 2050, including some subnational ones. Economic projections are being made to 2050 and even beyond. Up to 2050, they are based on qualitative scenarios of economic change. But beyond several decades, projection has been more in the domain of futurism than science—based on idea generation (e.g., Coates et al., 1997). Many social scientists question the quality of projections beyond 2050.

Wilbanks said that responses to the limited state of the science have included development of rich narrative "story lines," such as were created for the Millennium Ecosystem Assessment and for some economic projections to 2050; describing alternative futures of interest and working back from them with quantitative scenarios; and using participatory analytic-deliberative processes to generate qualitative descriptions of futures.

Wilbanks identified four key questions for the workshop:

1. What does the community need in order to generate mid- and long-term projections?
2. What is the current state of the science/art for such descriptions of the future?
3. How might the state of the field be improved, both in the short term as a basis for the IPCC Fifth Assessment, and in the longer term?
4. What suggestions can be offered for near-term action?

2

Needs for Socioeconomic Scenarios

IMPACTS, ADAPTATION, VULNERABILITY, AND IPCC WORKING GROUP 2

Christopher Field

Christopher Field, the leader of Working Group 2 (WG2) for the Fifth Assessment of the Intergovernmental Panel on Climate Change (IPCC), discussed scenarios in relation to this assessment. He said that instead of a list of possible impacts, the assessment needs to produce information about the possible future that will be useful for decisions. To achieve this goal, the assessment needs better integration of climate science and climate impacts in forms that help WG2 make good use of climate model outputs from Working Group 1 (WG1). The assessment also needs to put climate change in the context of other stresses within a consistent set of socioeconomic futures. He noted that there is some question about whether probabilities should be associated with the scenarios. What is important, he said, is to provide better treatment of extremes and disasters. Thus, the most important change in direction is probably to present issues in a way that provides a good foundation for decisions about risk, especially about low-probability, high-consequence events. The assessment also needs an expanded treatment of adaptation using a small enough set of scenarios to be useful. It also needs better integration of adaptation, mitigation, and development. Finally, the assessment is committed to the challenging task of developing a more comprehensive treatment of regional aspects of climate change. He identified two cross-cutting themes: (1) consistent evaluation of uncertainties and (2) better treatment of economic and noneconomic costs. He summarized by emphasizing that the IPCC Fifth Assessment needs to move from emphasizing the point that climate change is real to providing information that stakeholders need.

IPCC WORKING GROUP 3 PERSPECTIVES ON NEEDS FOR SOCIOECONOMIC SCENARIOS[1]

Ottmar Edenhofer

Ottmar Edenhofer, chair of IPCC Working Group 3 (WG3), presented a WG3 perspective on the scenario process, including coordination issues. He noted that the current outline of the WG3 report frames the issue in terms of risk and then examines pathways for mitigation by sector, including a chapter on human settlements, infrastructure, and spatial planning. WG3 will look at a number of transformation pathways developed by the scientific community. It is intended that the pathways will be explicit about unintended side-effects, such as leakage from carbon storage projects and effects of bioenergy development on food security, in order to show both the mitigation choices and their implications. Edenhofer said that, although the representative concentration pathways (RCPs) provide a minimum of consistency across the working groups, there is also a need for a realistic representation of the policy space that does not simply assume that all options are feasible.

His understanding is that the RCPs will be analyzed by the climate community to yield patterns of climate change. He said that the socioeconomic variables coming from the IAM community need to be downscaled, and the assessment needs to explore the full range of transformation pathways for each RCP. He suggested that it might be useful to develop what he called RSPs—representative socioeconomic pathways—which could be a basis for connection between the IAM and the IAV communities. He said that scenarios would need to identify demographic, economic, and other drivers and could serve as exogenous drivers for baseline conditions as well as for policy scenarios. He suggested that RSPs could be combined with policy scenarios, with each combination yielding an emissions trajectory. He also suggested that the process could also develop scenarios with "second-best" policies.

---

[1]Edenhofer's presentation is available at http://www7.nationalacademies.org/hdgc/ Mitigation_and_IPCC_WG_III_Presentation_by_Ottmar_Edenhofer.pdf [November 2010].

ECOSYSTEM SERVICES

Anthony Janetos

Anthony Janetos spoke about the need for scenarios to consider ecosystem services, which have not received much attention in past climate assessments, although he considered them important.[2] He noted that the concept is anthropocentric: it includes only products of ecosystems that benefit humans, but that have no cost until there is a need to replace them. He said that the Millennium Ecosystem Assessment (MEA) was the most complete effort to develop this idea in the past decade.[3] An activity on the same scale as IPCC but not sponsored by governments, the MEA produced volumes on current conditions and trends and a volume on scenarios for the future of ecosystem services that includes alternative policy choices. The report categorized services as supporting, provisioning (typically services that are priced), and cultural (e.g., esthetic). He said that the assessment was not an exercise in justifying ecosystem preservation. It recognized that some past changes in ecosystems were of positive value to humans but considered that this value may degrade in the future in ways that are not well reflected in typical economic accounting.

Janetos suggested that the IPCC should pay attention both to direct dependencies on ecosystems (e.g., for agriculture, fisheries, and water supplies) and to indirect dependencies (biological diversity, carbon sequestration). The latter is new territory for climate assessments. In addition, he said, the IPCC should capture differences in the demand for services (e.g., market versus subsistence demands, such as for fuelwood) and also address governance issues, such as the roles of resource management agencies, the private sector, and household decisions, as well as differences in governance between developed and developing countries and changes in governance over time.

Janetos said it will be important for the assessment not to try to monetize everything and also to use some natural units. He found the concept of the social cost of carbon problematic, noting that estimates are widely different because of the difficulties of monetizing all the ecosystem services. He said the community needs to find ways to merge economics-based and other forms of modeling. He also suggested that WG2 explore the relationships between the supply of services and the resilience of

---

[2]The concept of ecosystem services was developed to provide a way to place value on the ways in which ecological systems improve human welfare that are not captured in commercial markets (see Costanza et al., 1997).

[3]The MEA, which operated from 2001 to 2005, involved more than 1,360 experts worldwide in assessing the consequences of ecosystem change for human well-being. Information is available at http://www.maweb.org/en/index.aspx [November 2010].

societies. He said that the MEA did some of this, but the conclusions have not been very visible in policy discussions.

In a comment at the end of the presentation, Granger Morgan expressed the view that ecosystem services should not be the sole framing of ecological impacts. He noted that the U.S. Environmental Protection Agency now analyzes ecosystem impacts only in terms of monetized ecosystem services and that it is important to monitor and report impacts in natural units as well.

GLOBAL ENERGY ASSESSMENT[4]

Nebojsa Nakičenovič

Nebojsa Nakičenovič discussed scenarios as used in the Global Energy Assessment (GEA), a large nongovernmental analytical effort.[5] He said that the GEA may be one of the first assessments using an RSP approach. Its scenarios follow a simple logic based on indicators, such as universal access to energy by particular dates, and consider the effects of such variables. GEA produced three transformational scenarios describing energy access and other variables. The scenarios are based on a single counterfactual reference scenario. The scenarios use a single set of population and economic growth projections, all gridded. They vary greatly, for example, in degree of urbanization and, partly because of that and other salient drivers, in fuel mix. All three scenarios lead to the similar climate outcome of stabilizing the future global mean temperature increase to 2° Celsius. This is achieved through different patterns of change in energy systems. All of the scenarios include significant development of carbon capture and storage and expansion of zero-carbon energy sources, including renewables and nuclear energy. The scenarios describe futures with a lot of efficiency and lifestyle changes. He concluded by saying that what are needed are analyses that show different socioeconomic futures that lead to both similar and different outcomes.

---

[4]Nakičenovič's presentation is available at http://www7.nationalacademies.org/hdgc/Energy_Trends_and_Global_Energy_Assessment_Presentation_by_Nebojsa_Nakicenovic.pdf [November 2010].

[5]The Global Energy Assessment, based at the International Institute for Applied Systems Analysis, is a multiyear international effort to provide national governments and intergovernmental organizations with "technical support for the implementation of commitments aimed at mitigating climate change and sustainable consumption of resources." Information is available at http://www.iiasa.ac.at/Research/ENE/GEA/index.html [November 2010].

RELEVANCE OF THE NEW SCENARIO PROCESS

Richard H. Moss

Richard Moss gave an overview of the new scenario process developed by three research communities—integrated assessment, climate change, and impacts-adaptation-vulnerability. He noted that a paper on the process, of which he is lead author, will appear in the February 11 issue of *Nature* (Moss et al., 2010). Moss said that previously scenarios were prepared and used sequentially, from driving forces to narratives that produced emissions scenarios, which led to estimates of radiative forcing, which were fed into climate models and then to models of impacts. Because all this work takes time, impact estimates in one assessment were based on climate models that were contemporary with a previous assessment.

The IPCC decided in 2006 not to develop new emissions scenarios, thus prompting an interdisciplinary group of researchers to develop a new process to develop and apply consistent scenarios across the three distinct research communities. The new parallel process is intended to enhance coordination across these groups. Rather than starting with detailed socioeconomic scenarios, it starts with radiative forcing. A set of radiative forcing pathways was selected to map out a broad range of future climates. New climate and socioeconomic scenarios will be developed during the parallel phase of the new process. Some of the new socioeconomic scenarios will be directly related to the radiative forcing pathways (e.g., what socioeconomic paths lead to a particular pathway and level of forcing in 2100); some new socioeconomic scenario work will not be tied directly to the RCPs.

Moss reviewed expectations for three sets of products: the RCPs, the climate scenarios, and the socioeconomic scenarios. The four current RCPs were generated from the available IAMs and are intended for use in climate models. They are defined in terms of radiative forcing in 2100. The RCPs differ in the forcings they show and in the trajectory of forcing. Two sets of climate model scenarios will be developed. In one set, extending to 2100 (or to 2300 in some cases), runs will be conducted at 1- or 2-degree geographic resolution. A second set of higher resolution (0.5 to 1 degree) will provide 2035 "decadal predictions" and will be run off a single scenario (RCP4.5), thus allowing larger model ensembles, higher resolution, and presumably better information on natural variability and extreme events. There are debates about whether decadal predictions are skillful and useful. Moss noted that the climate modeling community has been careful about prioritizing work to facilitate intercomparison of models and suggested that the same should be done with socioeconomic scenarios—the focus of this workshop. He noted that they are needed to provide the context for interpreting changes in climate and for exploring the different policies, technologies, and other conditions that could be consistent with different climate futures. He noted the need to be clear about what one needs to project, tying the scenarios to a particular purpose, and that the IPCC needs to identify a manageable number of key cases for use in its assessment.

Moss concluded by expressing the excitement developing in this field and related it to integration across the communities. But he pointed out that progress will require resources. He said that user support is critical for further development of the scenarios, and there is a need to engage researchers in developing countries.

DISCUSSION

Several participants raised questions for discussion. One question was whether the IAM community is concerned about consistency in socioeconomic assumptions across scenarios and whether good progress could be made with a short list of socioeconomic variables. It was noted that in IAV analysis, qualitative approaches are used more frequently than quantitative ones. Following this idea, the question was raised as to whether it would be possible to have a small number of alternate visions of the societal future to work from and to relate to RCPs. Nakičenovič pointed out that there are already a number of accounts of the socioeconomic future in the MEA and elsewhere, so it would be possible to link some of the downscaled socioeconomic scenarios to climate and to the RCPs. Gary Yohe noted that the scenarios developed for the last IPCC process are still potentially useful. Edenhofer suggested that the RCPs could be a focal point for consistency purposes and that RSPs could also be a focal point for consistency across IAV and mitigation analyses (for example, there could be pathways characterized by high urbanization or by high economic growth).

Some participants emphasized the need to cluster the narratives to avoid a proliferation of too many scenarios. Many researchers think that the community should work with a small number of socioeconomic pathways, but that developing these so that they are compatible with the large number of potential uses for IAV and mitigation analysis is an important challenge.

Marc Levy proposed that RCP and RSP processes need not be similar. For RCPs, only the aggregate matters. Impacts, however, are highly varied regionally, which suggests that the process for producing socioeconomic pathways should not be modeled on the RCP process. He claimed that skill in projecting emissions was not necessarily correlated with skill in projecting socioeconomic conditions.

George Hurtt noted that coupled carbon-climate models need input data on spatially specific land use activities, as well as biophysical feed-backs from land use. He said that a new generation of fully coupled earth system models is now in development, with socioeconomic information included. He pointed out the need to relate this process to development of new socioeconomic scenarios in the new process.

Gerald Nelson asked whether current models yield useful policy advice on key questions, such as whether soil carbon is included in an offset or compensation regime.

There was some discussion as to whether probabilities will be asso-ciated with the scenarios, a subject of lively debate in the community. It may be that the probabilities of certain forcing pathways may be easier to estimate than the probabilities of socioeconomic futures.

3

Evolving Methods and Approaches

PHILOSOPHIES AND THE STATE OF SCIENCE IN PROJECTING LONG-TERM SOCIOECONOMIC CHANGE[1]

Robert Lempert

Robert Lempert spoke on various approaches to projecting long-term socioeconomic change and their effectiveness. He noted that scenarios can have several functions: to provide consistent inputs to analysis, to inform decisions, to transform world views, and to entertain. Their purposes include predicting the future, identifying what might happen, and identifying ways to reach goals. They may be exploratory, to identify and consider many possible futures; they may be intended for decision support; they can be formal or intuitive, simple or complex.

A small evaluative literature exists on scenarios for long-term decisions, including a study by RAND-Europe that looked at about 50 evaluative studies. Lempert said that the many available methods of projection derive from three schools: (1) the intuitive logics school, starting with the work of Herman Kahn, which begins with drivers and develops scenarios from them; (2) the La Prospective school (Godet, Berger), which emphasizes visioning and focuses on desired end states; and (3) a school of probabilistic modified trends, which uses expert elicitation to identify possible surprises. A variety of techniques for describing the future are

---

[1]Lempert's presentation is available at http://www7.nationalacademies.org/hdgc/ Philosophies_and_State_ of_Science_Presentation_by_Robert_Lempert.pdf [November 2010].

adopted, including expert judgment, backcasting, and various modeling methods.

Scenarios can produce a number of benefits. One impact is to overcome cognitive barriers (e.g., optimism biases, strategic use of uncertainty, ambiguity aversion, status quo bias). Scenarios use various mechanisms to overcome the barriers. For example, they can focus on possibilities rather than predictions. There is some evidence that scenarios can actually reduce overconfidence and increase the coherence of beliefs, and in one study with firms, the use of scenarios was correlated with future profits.

Challenges in the use of scenarios for climate analysis lie in (a) the potential for divergent views on what scenarios are, potentially leading to an illusion of communication; (b) the tension between the desire for consistency and the need to consider surprises (e.g., formal models tend to leave out the discontinuities); (c) the need to include context in scenarios (the trade-off between simplicity and utility, the tendency to ignore scenarios when they can't deal with the projected futures); and (d) the need to emphasize process over product in decision support (National Research Council, 2009a).

Scenarios for decision support can be framed as a way to analyze vulnerability under existing plans and response options. Stakeholders in a decision may disagree on much but still agree on the need to think through how and when an option may not work. A database of many model runs can help identify the key drivers of failure and the scenarios leading to failure, thus helping in the consideration of response options. For example, a group at RAND looked for climate scenarios that failed to reach a concentration target of 450 ppm and found that, in most of these cases, carbon capture and storage and transportation systems failed to meet their targets.

The literature indicates that vast arrays of scenario methods are used for many different purposes. Some empirical evidence exists on the factors affecting scenario effectiveness in various applications, and these studies emphasize the importance of process (rather than products) and of close coupling with decision makers as determinants of effectiveness. Lempert concluded that, for some purposes, people may want to think less about developing standard scenarios and narratives and more about developing tools that particular decision makers can use to identify multistressor vulnerabilities and to consider their decision options.

DEMOGRAPHIC CHANGE[2]

Thomas Buettner

Thomas Buettner spoke on projecting demographic change. He noted that even current population is uncertain: half the world's people have no vital records, and because decennial censuses are rare, most of the information on demographic change comes from sample surveys. And uncertainty about current conditions is a problem not only for low-income countries. Germany has not had a census for 20-30 years. Buettner said that in some countries, data collections are fragmented and driven by donor demands (e.g., the U.S. Agency for International Development). Moreover, spatial resolution is a problem.

Buettner said that, in the past, demographic transition theory has guided population projections successfully. Now, however, about 47 percent of the world's population has reached the end of the demographic transition, and demographers do not have a theory for what happens after that. One possible path is equilibrium; one is a sustained path below equilibrium. Buettner also noted that the transition is stalling in some low-income countries. The low-fertility, low-mortality equilibrium predicted by transition theory is elusive. United Nations' (UN's) projections still assume "due progress" on the transition. They include past shocks but not possible future shocks or significant contextual changes.

Buettner noted some long-term demographic trends, including population aging and the "demographic dividend" of large cohorts of young people. He noted that some low-income countries have low fertility, and that high-income countries have slowing gains in life expectancy. Expected population growth in the next 40 years will be largely urban and located in low-income countries. The UN will soon release projections to 2100. The medium variant has world population stabilizing at less than 9 billion, but the high and low scenarios are very different from that.

ECONOMIC DEVELOPMENT[3]

Gary Yohe

Gary Yohe spoke on the drivers of economic development and the ways economists look into the future. There are large unknowns, such as about when countries start to develop rapidly, how they will handle pol-

---

[2]Buettner's presentation is available at http://www7.nationalacademies.org/hdgc/ Demographic_Change_ Presentation_by_Thomas_Buettner.pdf [November 2010].

[3]Yohe's presentation is available at http://www7.nationalacademies.org/hdgc/ Economic_Development_Presentation_by_Gary_Yohe.pdf [November 2010].

lution, among others. The Special Report on Emissions Scenarios of the Intergovernmental Panel on Climate Change (Intergovernmental Panel on Climate Change, 2001) looked at the important drivers: per capita gross domestic product (GDP) and emissions, demography, institutional development, development patterns over time, international trade and development, innovation and technological change, the distribution of income and opportunity, and energy intensity over time. Economic projection models presume that capital investment drives economic growth. GDP is connected to emissions through parameters, including carbon intensity of GDP, which can change over time. The devil is in the details, especially at the regional level. One can build a growth model using parameters for capital, labor, and perhaps fossil and nonfossil energy. The shares in the energy sector may change with the relative prices of energy. The models assume that these changes depend only on the price of carbon, but there is a need to look at other drivers of change, as there may not be a price for carbon.

Yohe said that economics is not good at predicting inflection points. He also noted that business cycles are more moderate in higher income countries and that socioeconomic diversity implies diversity in development paths. For example, if capital-intensive technologies are placed in a low-income country, the result might be huge unemployment. He reflected on Rostow's analysis of the prerequisites for economic takeoff (enlarged demand for a sector and the possibility of producing in the sector, which generate capital for the leading sector, the development of which can spill over across the entire economy, leading to a rapid growth rate).

In the discussion of this presentation, Ottmar Edenhofer suggested that people who study endogenous growth, including technological change, should be included along with growth economists in developing economic scenarios. Granger Morgan asked whether the IPCC is effectively prohibited from considering certain unappealing scenarios (e.g., nuclear war, global pandemic, failure of development in some countries). Field replied that those prohibitions have existed in the past, and Buettner added that countries complain to the UN if the projections run counter the country's development plans. John Weyant noted that this workshop is outside the IPCC process, in part to allow for consideration of such possibilities, so analyses of them enter the literature and can be considered by the IPCC.

Edenhofer said that Working Group 3 will have a chapter on policies that will include global, national, and subnational ones. Nakičenovič noted that demographics are not independent of the other variables. For example, future migration to cities will depend on economic development paths. Anthony Janetos noted problems with distortions on data and cited measures of forest cover as an example. He said that countries with few trees have a more expansive definition of what a forest is and that physical modeling based on countries' reports of forest cover will be wrong.

Finally, Lempert asked if there are bounding constraints on rates of economic growth, on amplitude of the business cycle, and on other major economic parameters. Yohe said there are data on this, but noted that both the rate of growth in China over a long period and its quick recovery from the recent recession have surprised economists.

CONNECTING NARRATIVE STORY LINES WITH QUANTITATIVE SOCIOECONOMIC PROJECTIONS

Ritu Mathur

Ritu Mathur discussed issues and methodologies related to connecting narrative and quantitative projections. She noted that many socioeconomic conditions can be consistent with a single forcing pathway. Accordingly, various researchers and users may end up considering widely varying socioeconomic or even emission trajectories for a particular region for the same forcing pathway at the global scale. There can be wide variation in trajectories of emissions depending on whether assumptions regarding technological progress and consumption behavior are optimistic or pessimistic. For example, widely divergent pathways have been examined for India in various studies, and while some are due largely to differences in socioeconomic assumptions, some are related to differing perceptions about the pace of technological progress.

Mathur also discussed the use of backcasting approaches to examine low-carbon pathways across regions to arrive at a warranted global emission trajectory. In such studies, there is often a disconnect between the process of allocating emission reductions across regions in alternative scenarios and the application of a backcasting approach at the regional-local level to introduce emission reduction choices that can meet the predetermined levels. Moreover, there are issues in harmonizing global assumptions defining story lines with local conditions and resultant emission trajectories, since the processes are disjointed and do not always allow for reassessing the distribution of emissions across regions. This leads to difficulty in making bottom-up analyses meet the numbers in the regional and global models.

A study done by The Energy and Resources Institute jointly with the Oak Ridge National Laboratory (The Energy and Resources Institute, 2009) examined the potential impacts of relatively severe climate change on 11 states of Northern India. In this study, narratives were used to develop socioeconomic scenarios. These were based on four story lines demarcated on the basis of the relative importance attributed to environment and development and government versus market solutions. These narratives were then quantified with a focus on demographic, economic, and energy-related indicators, as well as sectoral indicators for water, agriculture, and health. These variables were quantified at the state level, taking into account decadal variations in the past and the qualitative story lines for the future.

Given the results of this study, Mathur concluded by asserting the need for further integrated assessment models and impact-adaptation-vulnerability analyses to generate more realistic and robust predictions on climate-related risks. This would require greater involvement of institutions at the regional and local levels to ensure that the assessed reduction potentials being considered in global studies allow for a better encapsulation of regional changes that are likely in the future.

QUANTITATIVE DOWNSCALING APPROACHES[4]

Tom Kram

Tom Kram made a presentation on behalf of Detlef Van Vuuren (who was unable to attend), based on an article in preparation by Van Vuuren and his colleagues on quantitative downscaling approaches. Kram noted that many methods of downscaling have been tried and that the chosen preferred method depends on purpose, coverage, resolution, and the availability of information. Downscaled data need to be consistent with both the larger and smaller scales, as well as internally. The article distinguishes four approaches: (1) algorithmic downscaling, which can be done (a) proportionally (assuming every unit at the smaller scale is equal), (b) by applying the change assumptions for the larger unit to the smaller units and assuming that they will converge toward the central estimate, or (c) by applying exogenous scenarios; (2) methods of intermediate complexity using simplified formulas that are calibrated differently for different subunits; (3) complex models that can be applied at a small scale; and (4) fully coupled physical-social models that use changes in the models for one year as inputs to the next year's estimates. Some methods can lead to problems, such as when growth rate data for Asia are applied to Singapore. Kram concluded by saying that although there have been bad experiences with socioeconomic downscaling the past, better data and more advanced algorithms are now available. He said that although many methods are available, for global applications, simple methods might be adequate.

---

[4]This presentation is available at http://www7.nationalacademies.org/hdgc/ Quantitative_Downscaling_ Approaches_ Presentation_by_Tom_Kram.pdf [November 2010].

4

Driving Forces and Critical Uncertainties in Adaptation, Vulnerability, and Mitigation

DRIVING FORCES AND CRITICAL UNCERTAINTIES IN SCENARIO CONSTRUCTION[1]

M. Granger Morgan

Granger Morgan spoke on the role of driving forces in scenarios. He began by stating the assumption that in principle people really do treat scenarios as though they were forecasts or projections, and he asked how much detail is really needed and whether people would use it if they had it. It is very difficult, for example, to find examples of anyone who really uses all the detail in the Special Report on Emissions Scenarios (SRES). Commenting that representative concentration pathways (RCPs) are a step in the right direction, he emphasized that people are not good at predicting the future of basic parameters, such as primary energy consumption. He suggested thinking about three things:

1. Model switching. Different models are of use for different parameters at different time scales. For example, computable general equilibrium models of the world's economy might be reasonably believable for a decade or so into the future, but running them out for a century and believing the result is not reasonable. For the far future, doing a bounding analysis may be all that is reasonable.

---

[1]Morgan's presentation is available at http://www7.nationalacademies.org/hdgc/ Importance_of_driving_ forces_Presentation_by_Granger_Morgan.pdf [November 2010].

Morgan said that he and David Keith have critiqued scenarios on various grounds: detailed story lines fixate people on those particular stories, whereas there are other ways to get to the same end point. Overconfidence is ubiquitous; putting probabilities on scenarios is problematic; and path dependencies—the order in which different changes happen—can make a huge difference. Evidence indicates that consensus processes tend to understate the level of uncertainty, as indicated by individual experts' judgments. The literature produces a range that is narrower than the judgments of the individual experts.

2. Bounding analysis. An alternative to story lines is to ask people to suggest all the conditions that could lead to very high and very low values of a parameter of interest and have the list reviewed by experts to cull out infeasible conditions and suggest how the extreme values might come to pass.

3. Working the problem backward. If one propagates probability distributions down the causal chain, the uncertainties get huge. An alternative is to go backward on the chain, from outcomes of interest to the paths that could produce them. What possible outcomes do people most care about, and how could they get there? Which things would be problems that need to be addressed and avoided? People have trouble with this method. Research has found stakeholders refusing to do it, because they saw it as unwarranted speculation about bad outcomes.

Morgan said that methodological uncertainties in scenarios are often not appreciated. For example, time preference is modeled in ways that do not correspond to what people do, and important feedbacks in physical or socioeconomic systems are sometimes not taken into account. The uncertainties in the methodology can be larger than those that apply to the technical issues. Also, assumptions about who makes the decisions are sources of uncertainty.

BRIEF PRESENTATIONS ON SPECIFIC DRIVERS

Brian O'Neill spoke about the relationships of demographic drivers and emissions on climate outcomes.[2] First, he noted that the relationships between story lines and demographic drivers are often thought to be much more solid than they are. Second, emission scenarios can be consistent with a wide range of demographic drivers. The A2 SRES

---

[2]The presentation is available at http://www7.nationalacademies.org/hdgc/Population_Presentation_by_Brian_ONeill.pdf [November 2010].

scenario's population assumption could as well have been a lot lower, based on O'Neill's analysis of all the story lines that could produce the same results. One can have quite a lot of variation in emissions for a wide range of possible population levels, with the possible exception that high population may not be consistent with low emissions. His conclusion was that, although demography does matter to emissions, a full range of demographic changes is consistent with a wide range of emission pathways.

Gary Yohe spoke about socioeconomic drivers of impacts, adaptation, and vulnerability (IAV).[3] He emphasized that IAV is site-specific and path-dependent. He said it is desirable to preserve some degree of internal consistency of local exposure and sensitivity with global portraits and to explore complementarity or conflict between IAV futures and mitigation futures. However, he noted that it is difficult to describe the drivers of the capacity to mitigate or adapt, and it is even harder to describe the link from capacity to action.

Yohe identified these determinants of adaptive capacity: availability of options; availability and distribution of resources; human and social capital, governance responsibility, and authority; ability to separate signal from noise; and access to risk-spreading mechanisms. He said that economic models assume the spreading of risk. This list is similar to lists in other fields (e.g., precursors for disease prevention, prerequisites for sustainable development conditions for efficient working of markets). Knowledge is still inadequate, although, for prediction of adaptations or evaluation of efficacy.

He noted that there is never a unique sustainable pathway. Internal consistency is important in scenarios, but there are many ways to get to a single end point. Results should be as quantitative as possible and linked to consistent global futures and socioeconomic and climate trajectories. He noted that, even after concentrations peak and temperatures stabilize, some outcomes will continue to evolve.

Nebojsa Nakičenovič spoke about technology as a driver. Assumptions about technology are fundamental to the outcomes of scenarios. He noted that technologies have changed lifestyles dramatically several times since the Industrial Revolution and that this is one of the biggest uncertainties there is. He cited Paul Raskin's argument that technology is not fundamental but a derived driver of change that reflects transformations in culture, governance, etc. He said that the greatest impacts will depend on change in places with high population and low current levels of development, suggesting that it might be useful to map energy access statistics

---

[3]The presentation is available at http://www7.nationalacademies.org/hdgc/Economy_ and_Infrastructure_Presentation_by_Gary_Yohe.pdf [November 2010].

against population density. Technological learning, in terms of cost of a unit of energy capacity, is often very significant, with costs per unit falling by one or two orders of magnitude. The important changes are not in individual technologies, but in the system, in which the convergence of technologies is a major unknown. He noted that in very low emission scenarios, many energy technologies are treated as though they all will be extremely successful in their evolution, although the real world will be considerably different and messier. He concluded by noting two big gaps in knowledge and analysis: technological change on the adaptation side and the relationships between change in mitigation technology and in adaptation technology.

Michael Replogle spoke about the transport sector and regional planning.[4] This is perhaps the fastest growing area of greenhouse gas (GHG) emissions, now accounting for about 20 percent of emissions and growing fastest in countries outside the Organisation for Economic Co-operation and Development, particularly India and China. Replogle said that, in those countries, increased efficiency can offset growth in distance traveled. The context for growth in emissions from transport is rapid urbanization and urban income disparity: half the people in the world are still unable to afford vehicles, but vehicle ownership is closely associated with income, although there are differences due to policy and among continents. Most transportation analysis has focused almost entirely on technology strategies; little attention has been paid to strategies to avoid lock-in of motorization, even though it would produce huge benefits and huge negative costs per ton of carbon abatement. Replogle noted that GHG intensity per household is much lower in walkable neighborhoods, indicating that there are high and low carbon paths for urbanization, depending on subsidies for car use, road building, etc. The paradigm for mitigating emissions is to avoid and shorten vehicle trips, shift to more efficient modes, and improve the efficiency of each mode.

He identified a series of key driving forces of change in transportation GHG emissions: the pattern of urbanization; growing mobility related to globalization; technological innovation (including the availability of lower carbon vehicles and fuels and advances in information, communication, and automation); energy security and prices; transport security and terrorism; public health and safety issues (e.g., changes in physical activity, air pollution, and traffic accidents); congestion, livability, and economic competitiveness issues; infrastructure and energy finance; and climate change and adaptation, including dislocation and infrastructure

---

[4]The presentation is available at http://www7.nationalacademies.org/hdgc/ Transportation_Including_Regional_Planning_Presation_by_Michael_Replogle.pd [November 2010].

loss. He noted that, in some countries, increased technical efficiency has been used to increase vehicle size and power rather than to lower emissions per distance.

Replogle concluded by pointing out that market failures in transport abound (e.g., in the United States, free parking, $300 billion in motor fuel subsidies, fixed cost insurance pricing). The recent *Moving Cooler* study (Cambridge Systematics, 2009) concluded that the United States could reduce emissions by 25-40 percent at negative cost per ton, but he questioned whether its governance structures could move in that direction, for example, by following the Singapore model of urbanization and implementing user charges to reduce congestion and regulate vehicles for high performance.

Frans Berkhout spoke about policy and institutions, which he defined as rules for collective action. Institutions embody ideas about what is right and codify practices. Berkhout said that there is a need to include political scientists, sociologists, and anthropologists in the Intergovernmental Panel on Climate Change (IPCC) to get these issues considered. Political scientists tend to avoid making predictions, and they are reinforced in this by prominent political surprises, such as the end of the cold war. He noted that putting policy into scenarios presents a dilemma, because the scenarios influence policy. Policy makers often insist on policy-poor scenarios because they want to choose the policies, but policy-poor scenarios look unrealistic. He noted that although policy institutions resist change, there is also huge global institutional diversity. There will be democracies and dictatorships, for example, but systems change from one form to the other at surprising times. There are also different national systems of innovation, but each results from its own history in ways people are unable to predict.

He identified four drivers of institutional change: (1) the relative roles of governments and of businesses and other actors in governance; (2) the roles of knowledge, transparency, and accountability; (3) the interdependence of governance systems, including the possibility of learning across places; and (4) cultural norms on the value of nature and attitudes to risk. Each of these things is fairly stable but does undergo change, sometimes suddenly, for reasons that are not well understood. For example, novel institutional forms, such as carbon markets, do emerge. He suggested that the fifth IPCC report could bring expertise on institutions into the process. In a brief discussion, Ferenc Toth noted that climate vulnerability can vary a lot with the same level of economic activity, depending on institutional structure. As an example, he noted the adaptability of farmers in Thailand, who have a high level of autonomy in decision making.

Elmar Kriegler spoke about first- and second-best policies.[5] A first-best policy removes all market failures, resulting in an economically optimal outcome. A second-best policy optimizes the outcome given the constraint that one (or some) of the market constraints or failures cannot be removed. In the integrated assessment modeling community, the idea of second-best policy is used in a slightly more general sense, to address not only market failures, but also limitations to technological change and adoption and other constraints. For example, if carbon capture and storage is taken off the table because of public opposition or an accident, what would the best policy be? What effect would limited or fragmented participation of some countries have on reaching climate targets? Kriegler presented results of some modeling studies of the costs and capability of reaching certain emissions targets if certain technologies are taken off the table. He noted that there are also market failures in adaptation. For example, there is underprovision of public goods, such as flood protection.

Habiba Gitay spoke on ecosystem and water resource issues. She emphasized that drivers are linked, that single drivers can have many impacts, and that many of the drivers interact with climate change in non-linear ways. Moreover, the ways they affect ecosystems are place-specific. The drivers of change in these systems are also affected by development pathways. The IPCC in the past has not incorporated ecosystem perspectives well. Gitay presented a list of drivers of ecosystem change: urbanization and coastal settlement, including water withdrawals; land use and land cover change; trade, including in cash crops; population distribution; conflict; migration and urbanization; technological changes; policy changes; behavioral change, including values, choices, and resource consumption patterns; and, finally, climate change itself.

Gerald Nelson spoke on food, nutrition, and bioenergy issues. He said that increases in population of perhaps 50 percent and increases in income will imply a need for 70-100 percent more food on limited additional arable land. Agriculture also has a significant role in GHG emissions, accounting for 30 percent of emissions if forests are included. The mix of emissions matters. Increasing food production with the same ways of using nitrogen will increase GHG emissions, so there is a need to look to agricultural technology for managing emissions as well as for adaptation. He pointed out that development economists think of land the way they do capital and labor, as resources that can be added to, but that assumption is hitting its limits with respect to land. If land cannot be expanded, the analysis becomes more complicated, and the story changes. Also, the effect of technology is different for the different factors of production. The

---

[5]The presentation is available at http://www7.nationalacademies.org/hdgc/First_and_Second_Best_Policies_Presentation_by_Elmar_Kriegler.pdf [November 2010].

private sector has a good incentive to improve the productivity of capital and, to some extent, of labor. But for land and water, it can be much more difficult to capture the returns to improved technology. Agriculturalists are incredibly adaptive, but some crop models give scary results about the effects of temperature increases when they include limits to adaptation in the form of crop growth. Nelson mentioned his recent chapter on biofuels, which argues that they are not a good use of sunlight (Nelson, 2010). He said that nutritionists are rarely concerned about the effects of climate change on nutrition and proposed that this is worth further examination. He concluded with a strong statement about the need for improved data. For example, climate models don't provide information that is critical for the analysis of agriculture, such as about change in minimum nighttime temperature.

Kristie Ebi spoke about health issues. She began by saying that, after age 5, the risk of death from chronic disease is the same worldwide. The big differences in life expectancy are from communicable diseases—upper respiratory infections, diarrheal diseases, malnutrition, and others. She said that these problems are intractable in some areas, and long-term change in gross domestic product (GDP) won't necessarily solve them. There has been more than a century of malaria control in parts of Africa, with some short-term and local gains but long-term, systemic failure. These things are climate-sensitive, and the health effects mainly involve children. This does not change demography much, but malaria reduces GDP by about 10 percent in Africa.

Ebi said that the consequences for health of changes in ecosystem services, water, and agriculture can be noticeable, but it is difficult to tease climate effects out from interacting multiple stresses. The best predictors of life expectancy are simple medical care, education, and access to clean water and adaptation—not GDP. How well countries do depends on the responses of institutions, including the World Health Organization and private nongovernmental organizations. She said that the health sector does not use scenarios, because they don't show that the world is not reaching the Millennium Development Goals. The scenarios have not been realistic about the health sector. Public health is very different from health care delivery. She noted that in the Soviet Union, public health failed before the country dissolved; by contrast, Cuba has some of the best health care in the world.

DISCUSSION

The following issues were raised in the discussion:

- interactions between mitigation and adaptation in agriculture (e.g., holding more carbon in the soil leads to better water retention, which increases productivity and diversifies sources of income);
- a need for serious thought to very different worlds with the same radiative forcing, such as one with the entire biosphere managed for biofuels, food, etc., and one in which biosphere performs more traditional ecological roles;
- the importance of the difficult-to-monetize quantities;
- large expenditures on cultivars in Africa without investments on agricultural extension to implement them;
- the need to construct adaptation scenarios in a multisector context (e.g., Can goals for technologically driven increases in crop yields be reached given expected changes in climate and flood regimes?); and
- the need to increase sophistication of scenarios related to adaptation to match the sophistication of scenarios leading to mitigation.

The end of the afternoon was devoted to breakout discussions focused on long-term scenarios for adaptation and mitigation that could be used by the IPCC, the U.S. National Assessment, and other analyses.

5

Representative Concentration Pathways and Socioeconomic Scenarios and Narratives

John Weyant introduced the session, which was a discussion of representative concentration pathways (RCPs) from the perspective of the integrated assessment modeling (IAM) community. He said that the IAM community and individual IAM teams offer some frames for socioeconomic scenarios. Expressing concern that people outside the community would expect perfect synchrony among efforts, he suggested that a more appropriate standard of comparison is with past efforts. He noted that the workshop has been characterized by constructive comment.

CHARACTERISTICS, USES, AND LIMITS OF REPRESENTATIVE CONCENTRATION PATHWAYS[1]

Jae Edmonds

Jae Edmonds said that the RCPs were developed to quickly deliver emissions data to the climate modeling community. Taken from the open literature, they were to provide a selection of pathways leading to different degrees of climate forcing to be used as inputs to climate models, story lines, and impact, adaptation, and vulnerability (IAV) analysis. The RCPs are presented in radiative forcing units (w/m^2) from greenhouse

---

[1]Edmonds's presentation is available at http://www7.nationalacademies.org/hdgc/ Characteristics_Uses_ and_Limits_of_the_RCPs_Presentation_by_Jae_Edmonds.pdf [November 2010].

gases (GHGs) and aerosols but not including forcing from albedo changes or from mineral dust. The four RCPs are known by their levels of forcing in 2100: 8.5, 6, 4.5, and 3 (in w/m^2). Each was developed by a different research group using a different model, and thus the socioeconomic scenarios do not constitute a set, for example, with population or GDP of the highest RCP being the highest for those variables when compared with the other RCPs. All will start from the same historical baseline in 2000 (RCP 6 was still being harmonized at the time of the workshop).[2]

The RCPs are designed for climate modelers and therefore include the full suite of relevant gases, aerosols, and land use and land cover. They are downscaled to 0.5 degree, and scenarios are extended in a stylized way to 2300 to allow for climate model research on equilibrium behavior of the climate system. They are consistent with data back to 1850 for gases and land cover to 1700. They have data for over a dozen sectors. Edmonds emphasized that the RCPs were selected to bound a wide range of possible future forcing characteristics over time, not to bound socioeconomic uncertainty.

Emissions trajectories for the RCPs are openly available, but the underlying socioeconomic data are not yet available. Edmonds said that some in the community want to make more data available than went into the RCPs, so as not to overemphasize the RCPs. The research teams are trying to create additional socioeconomic scenarios, called replication ensembles, which would yield the same end states that their models produced. Noting that the drivers are not all downscaled in the RCPs, Edmonds emphasized that many different socioeconomic scenarios are consistent with any of these levels of forcing. Even the 2.6 results can be reached from any socioeconomic scenario—if the requisite policies are adopted.

During the parallel phase of the process, while the climate modeling groups are conducting climate model experiments, the IAM community is producing new socioeconomic scenarios, with alternative backgrounds, different technology availability regimes, and alternative shapes of the emissions pathways leading to the same end points—including even 2.6, which can be produced in various ways. An open question is what range of socioeconomic scenarios should be explored in the IAMs. It may be that the 8.5 scenario requires a more tightly specified socioeconomic future than the 2.6 scenario does.

Edmonds concluded by noting that the new scenarios process could provide an embarrassment of riches—hundreds of scenarios with different combinations of socioeconomic and climate changes. The RCPs

---

[2]They are documented in Moss et al. (2010), and data can be found at http://www.iiasa. ac.at/web-apps/tnt/RcpDb/ [November 2010].

provide detail down to 0.5 degree for drivers and are well documented. But because they come from four different models, they do not share a common set of reference assumptions from a socioeconomic perspective. There are socioeconomic circumstances that some models do not cover, and, in some respects, they are incompatible. For example, the 4.5 RCP has an increase in forests; the 2.6 RCP has an increase in crops and pastures. In the models, technology for crop productivity has as much effect on climate as energy technology, through effects on land use.

In the discussion, Gary Yohe suggested that if there are many models with many drivers, a decision analyst could determine which drivers are most important to influence, in order to affect forcing. Edmonds pointed out that models would have to be comparable to be used in that way. Anthony Janetos noted that no one has yet investigated whether downscaled climate models are consistent with multiple socioeconomic scenarios at the same scale.

MULTIMODEL ANALYSIS OF KEY ASSUMPTIONS UNDERLYING REPRESENTATIVE CONCENTRATION PATHWAYS

Tom Kram

Tom Kram presented a quantitative comparison of the RCPs prepared by Detlef Van Vuuren, who could not be present. Each RCP is internally consistent, and each could be used to drive climate models as a basis for impact assessment (if information is included to indicate exposure and other impact-related variables) and for mitigation analysis (again, with additional information, such as baselines, targets, and assumptions about technology and governance).

A crucial question for creating additional scenarios is how strongly assumptions about socioeconomic change are correlated with outcomes. Kram examined this question by comparing scenarios, adding in newly published material and material from the RCP groups. The results have five implications: (1) There is very little correlation between population assumptions and radiative forcing—any reasonable population scenario could coexist with almost any of the emission outcomes. (2) The full range of estimates of gross domestic product (GDP) is consistent with all the end points for forcing, except that the 2.6 scenario was consistent only with the lower part of the GDP range. (3) Primary energy consumption is related to emissions, in that the only scenarios that result in 8.5 w/m^2 involve burning large amounts of fossil fuels. (4) CO_2 emissions are closely correlated with radiative forcing—the 2.6 scenario requires net carbon storage by the end of the century. (5) Forest cover varies considerably across the models more than with radiative forcing level, suggesting that this result is sensitive to model assumptions. Together, the models do not cover all the plausible futures.

Kram said that, in order to use the scenarios to study IAV questions, it would be necessary to map levels of vulnerability against the climate signal given by the scenarios. It might be very useful to analyze two or three levels of vulnerability against several strengths of climate signal. A different approach is needed for mitigation, Kram said, and the mitigation community may want to examine whether particular RCPs are consistent with certain assumptions about technology or about potential international agreements.

In summary, Kram said that these RCPs, especially the middle ones, are consistent with a very broad range of futures in terms of key socioeconomic variables. This suggests that looking at a small set of socioeconomic scenarios might be very useful in IAV research.

In the discussion, the following issues were raised:

- The most recent scenarios tend not to cover the full socioeconomic space. For example, most of the estimates of population are near the UN median projection.
- The conclusions about low correlations are based only on a first pass and reflect a selection bias in the scenarios and the lack of examination of interactions in which some combinations of economic assumptions are inconsistent with some pathways.
- There is debate in the community about whether any 2.6 scenario is reasonable. That scenario was created as an "only-if" scenario—the idea was that 2.6 could be reached only if all the assumptions in the scenario are met.
- The potential to change net emissions with agricultural technology and increased uptake of carbon in soils may not have been fully examined in all the models.
- Interaction between the IAM and IAV communities will be needed to consider interaction effects and to develop skeletally described scenarios for the RCP forcing levels for the IAV community. Such analyses would need conceptual design, for example, to decide on how to focus on particular vulnerable groups.
- An econometric approach might be used to capture interactions.
- Many scenarios cannot reach 2.6. It might be useful to develop a 3.7 scenario—a value that is in a lot of other analyses. Second-best solutions are important for reaching 3.7.
- Is it sensible to put probabilities on the scenarios? This has been tried with simpler models, for example, using multimodel analysis of key assumptions underlying the RCPs Monte Carlo analysis.

6

Lessons from Experience

PERSONAL EXPERIENCES WITH SCENARIOS

Peter Schwartz

Peter Schwartz spoke from personal experience with doing scenarios since 1972 and climate scenarios since 1977, when he was looking at global cooling. In his view, a scenario is a tool for ordering perceptions about alternative futures in which one's decisions are played out. Its purpose is to improve decision making today, not to get an accurate prediction of the future. The test of success is not in getting the prediction right, but in improving action now. For example, in the 1980s, an oil company asked whether there is a scenario in which the cold war ended. Schwartz's analysis concluded that there was, and that changed the company's behavior.

Schwartz said the key to effective scenario development is to get the question right. They use the acronym STEEP—for social, technological, economic, environmental, and political—to indicate that these factors interrelate, but all these driving forces do not relate neatly to each other. He compared scenario development to rocket science. "I'm a rocket scientist. This ain't rocket science. Rocket science is easier. You have reliable equations in rocket science." He said that experience matters, warning that people are deceiving themselves if they use probabilities, because attending to probabilities leads to inattention to the long tails and to being surprised—an outcome to be avoided.

Schwartz emphasized the need to distinguish things that are predetermined from critical uncertainties that make a large difference to long-term outcomes. There are some predetermined elements in 30-50 year time frames, but not on longer scales. Even demography can change fundamentally in a decade. Even things considered physically impossible can become possible in a decade. The task in modeling, Schwartz emphasized, is to have different models that embody different theories of change—not different runs of the same model, which is sensitivity analysis. Rigorous models are needed that lead to different conclusions—different interpretations of history, leading to different projections.

Building models thus requires both rigor and imagination, even for a 50-year time frame. People are remarkably capable of failing to see signals of impending change. Good scenario thinking allows one to see the signals of fundamental change when they arrive. Schwartz concluded by warning that people need to take the idea of story telling seriously, because stories give meaning to facts.

THE U.S. NATIONAL ASSESSMENT[1]

Thomas Wilbanks

Thomas Wilbanks spoke about lessons from the U.S. National Assessment (USNA). The first USNA occurred in 1997-2000, with $14 million of federal funding.[2] It included 16 regional and 5 sectoral assessments and a top-down overall document. Wilbanks noted that a survey-based effort led by Granger Morgan to gather lessons from the first USNA found that many participants thought the future would be like the present except for the climate, and others thought that nothing useful could be said about the socioeconomic future because it is so complicated (Morgan et al., 2005). Morgan's group produced a guidance document, asking each group to pick one or two factors that they thought would be influential and to consider different values of these; however, this guidance was little used. The survey study concluded that the process got considerable stakeholder involvement and that those who were heavily involved were more favorable about the process. It concluded that the USNA worked best with the ecosystem topic, which was done very well. People considering the USNA concluded that it was very imperfect, but that a better process would probably have reached the same conclusions.

The initial vision of the first USNA was bottom-up; a top-down part

---

[1]The slide presentation, which was prepared primarily by Granger Morgan, is available at http://www7.nationalacademies.org/hdgc/US_National_Assessment_Presentation_by_Granger_Morgan.pdf [November 2010].

[2]Reports of the U.S. National Assessment completed in 2000 can be found at http://www.usgcrp.gov/usgcrp/nacc/ [November 2010].

was added during the process, and there was tension about this. Local people found the top-down scenarios not to be dealing with their main concerns. Concerns were also expressed about how to integrate different assessments that were conducted in different ways and funded at very different levels. The USNA process showed that stakeholders can contribute useful content and perspectives and that the process can build communities. Decisions about how to do an assessment and whom to involve need be anchored in a shared idea of why. There needs to be consistent funding and consistent guidance, and there is the challenge of reconciling the need for general conclusions with specific needs.

In discussion, it was asked how scenarios might have improved the USNA process. Wilbanks said that there were ideas of socioeconomic futures in the assessments, but not the models. The socioeconomic projections were introduced late and rarely used. It was also suggested that projections from the government often lacked credibility. For example, government projections of immigration assume that the law is obeyed.

Kathy Jacobs spoke about planning for the next USNA, which is due in 2013. The first regional meeting will be held in Chicago to build a plan for this year. The National Oceanic and Atmospheric Administration will support 6-10 regional workshops to get input on design of the USNA. The agencies want to learn lessons from the past, but for many of them budgets are tight.

It is intended that the USNA will assess capacity for adaptation and mitigation, not just examine impacts as before. Also, many more people will be involved than before. The agencies intend to work out a way to structure a long-term program, not just a single report. In fact, drafting the required report could be seen as a distraction from creating a long-term process. The goals include engaging agencies beyond the U.S. Global Change Research Program and engaging stakeholders. It is important to ask the right questions and to coordinate across regions, sectors, and scales. It will also be necessary to draw clear boundaries between the USNA and the national adaptation strategy, which is expected to deliver a report in September 2010. It is intended that scenarios will be an important part of the process, but it is not yet clear how they will interface with the entire assessment. Jacobs advocated looking at the interaction of regions and sectors; of energy, water, and agriculture; and so forth. She wants more investment in scenarios and in data related to the key issues. She noted that the assessments report of the National Research Council (2007) suggested conducting a broad, blanket assessment that would try to anticipate bad outcomes and that investments should be made in science related to avoiding those outcomes.

Several points were raised in discussion:

- In response to a question, Jacobs said that national security would be included in the USNA.
- Jacobs said she would like the USNA to relate to the Intergovernmental Panel on Climate Change (IPCC). For example, material prepared for the USNA by June 2013 could be cited in the next IPCC assessment.
- An interactive support capacity could be set up to help people interpret and apply scenarios. Jacobs noted that, in the past, the USNA provided generic data without asking what was needed.

THE MILLENNIUM ECOSYSTEM ASSESSMENT[3]

Gerald Nelson

Gerald Nelson talked about the Millennium Ecosystem Assessment (MEA), based in part on slides from Detlef Van Vuuren. Focal questions were the consequences of 50 years of development for ecosystems and then for human well-being. The MEA group strongly supported a story line approach and was skeptical about quantitative models; quantitative modeling was decoupled from story line development. Nelson said that ecosystem analysts had trouble sharply defining the boundaries of an ecosystem.

The MEA described four worlds, considered equally plausible, called (1) Global Orchestration, (2) Order from Strength, (3) TechnoGarden, and (3) Adapting Mosaic. It used integrated assessment models that had never communicated with each other before. There was little exposure to scenarios and serious jargon problems, some going undiscovered for years. It was easy for participants to talk, more difficult to write scenarios, and even more difficult to quantify them. Nelson concluded that the MEA has had some impacts. For example, the term "ecosystem services" is used more (e.g., in Congress), and there are efforts to pay for these services. There is also increased recognition of the need for agricultural and socioeconomic sciences to talk to each other. However, people quote the MEA as having made policy pronouncements that it did not make.

In discussion, it was suggested that the focus on story line narratives and modeling in the MEA was inappropriate. One participant commented that if the main focus was on current impacts and trends, it is no wonder that future scenarios were not a major focus. He suggested that the MEA could have built on hypotheses about end points as a way to focus discussion on how to get there.

---

[3]The presentation is available at http://www7.nationalacademies.org/hdgc/Millennium_Ecosystem_ Assessment_Gerald_Nelson.pdf [November 2010].

THE ASIA LOW-CARBON SOCIETY PROJECT[4]

Mikiko Kainuma

Mikiko Kainuma described this project, which started 6 years ago and focused on 2050, looking at compliance by each country with its own targets.[5] It has been analyzing what is needed to reach goals using a backcasting approach. The Low-Carbon Society project considered the relationship of each of a long list of issues to the desired state. The study identified 12 actions that would make possible emissions reductions of 70 percent in Japan, as well as the barriers for each. It included the design of a 2050 house and visualization techniques for a low-carbon society. The Japanese government is considering these actions.

The project concluded that early investment can reduce cost and enhance energy efficiency, but funds are needed. Technology options have been analyzed for Japan, India, and China. The project has also developed scenarios for Asian cities. Kainuma said that low-carbon visions are needed and that scenarios provide visions and focus on needs (knowledge, institutions, poverty reduction, etc.) but that political leadership is also essential.

THE IPCC SPECIAL REPORT ON EMISSIONS SCENARIOS

Volker Krey

Volker Krey spoke briefly about the Special Report on Emissions Scenarios (SRES) process on behalf of Nakičenovič and himself. He noted that SRES uses "proxy" drivers, such as population, but does not address "ultimate" drivers, such as values and needs, power structure, and culture. He summarized the ways the SRES scenarios addressed such drivers as population, economic growth, and technological innovation in developing emissions paths. He noted that many of the recent scenarios have population values close to the middle range of the earlier SRES scenarios, so they span a smaller range of values.

---

[4]Kainuma's presentation is available at http://www7.nationalacademies.org/hdgc/Asia_Low-Carbon_Society_Project_Presentation_by_Mikiko_Kainuma.pdf [November 2010].

[5]For further information on this project, see http://www.lcs2050.com/ [November 2010].

THE UK CLIMATE IMPACTS PROGRAM AND
THE NETHERLANDS EXPERIENCE[6]

Tom Kram

Tom Kram spoke from Frans Berkhout's slides on the experience of the United Kingdom and the Netherlands with scenarios. Started in 1999, the socioeconomic scenarios of the UK Climate Impacts Program (UKCIP) used high-resolution climate modeling and a range of local socioeconomic case studies to identify vulnerable areas.[7] The approach was like that of SRES, with schematic story lines developed from a few base parameters. There was also flexibility to tailor scenarios to meet needs.

In the Netherlands, the scenario effort focused on impact, adaptation, and vulnerability issues, to determine whether a few socioeconomic scenarios should be used to guide many programs in the country. The effort began with global, then European, and then Dutch scenarios on growth, trade, and so forth. They were merged with a fine-scale, spatially explicit model (possibly 100 meters), to show population changes, land uses, and implications for adaptation. In the future, the researchers want to simplify the models, include feedbacks, examine discontinuities, and develop visualizations.

DISCUSSION

In the discussion, it was noted that these examples use socioeconomic scenarios in different ways, such as to explore alternative futures (in the MEA) or to show ways to get to certain futures (IPCC and UKCIP). In some cases, they are relatively unused compared with climate scenarios (the USNA). Weyant noted that a model cannot be assessed without a question in mind, and it is useful to consider the different purposes. Sometimes scenario exercises identify where new research is needed. Janetos commented that, in the MEA, pairing natural and social scientists as authors of each chapter in the "conditions and trends" volume was very fruitful when it worked well.

Brian O'Neill emphasized the use of scenarios for considering the trade-offs among climate change futures, for example, efforts at mitigation and adaptation. He suggested that there is a need for socioeconomic

---

[6]The presentation is available http://www7.nationalacademies.org/hdgc/Quantitative_Downscaling_Approaches_Presentation_by_Tom_Kram.pdf [November 2010].

[7]The UKCIP develops climate projections and uses the resulting information to help organizations in the United Kingdom adapt to "inevitable climate change." For further information, see http://www.ukcip.org.uk [November 2010]./

scenarios for each RCP. Representative socioeconomic paths could be described on two axes: levels of vulnerability and adaptive capacity on one, and factors that make it easy or hard to mitigate (mitigation capacity) on the other. Socioeconomic factors could make it easy or hard to adapt or mitigate, giving a 2 × 2 matrix. A set of socioeconomic scenarios could be tested against all the RCPs, with research focused on whether they are consistent with each other.

Richard Moss supported Schwartz's idea of thinking about decisions that people can make in the context of scenarios that bound broad visions of the socioeconomic future.

Stephane Hallegatte noted that, in France, political pressures restricted the scenarios that could be presented, resulting in politically shaped decisions to look only at vulnerability to the current climate and to exclude impacts on ski resorts. Based on this experience, he wondered whether scenarios will actually be used. Others commented that many people (including decision makers) misunderstand visualizations as predictions and have difficulty making sense of scenarios. In addition, there is a need to consider many different kinds of vulnerability.

7

Reports from Breakout Groups

The participants divided into three breakout groups on the first day and two groups on the second day. The following summaries were presented by rapporteurs drawn from among the participants in each breakout group.

IMPACT, ADAPTATION, AND VULNERABILITY
SCENARIOS FOR 2020-2050

Kristie Ebi, Rapporteur

Kristie Ebi identified five major themes that arose in the discussion:

1. Climate is unlikely to be a major driver in most regions and sectors in this time period, so it is not possible to distinguish meaningfully among scenarios. It will be important to focus on extreme events and other particular events that might be important, making narratives consistent with actual development pathways—for example, it might be useful to talk in terms of the Millennium Development Goals, which she noted are likely to be unmet in any event. Ebi suggested that nonclimate events will be more important than climate in the short term.
2. The number of driving forces is too large to consider in a narrative. Thus, scenarios should focus on a few key drivers, such as governance and institutions, access to public-sector services,

ecosystem services, urbanization, and globalization and trade, and they should take a livelihoods approach to development pathways in certain regions and sectors.

3. There is a need for a few narratives as a framework onto which additional narratives can be added for particular sectors and regions. It is possible to analyze regions and sectors individually only for low degrees of climate change.

4. There is a need for baseline and policy narratives: no-regrets options, options with regrets, and overshoot situations.

5. There is a need to identify desired end points for this time frame (i.e., where societies would like to be) and develop scenarios by working backward from those states.

One group member added the need to distinguish global from local scenarios.

IMPACT, ADAPTATION, AND VULNERABILITY ISSUES TO 2100

Ferenc Toth, Rapporteur

Ferenc Toth identified the following themes from the discussion:

- It is important in this time frame to examine factors with large inertia, which change mainly in the long term, such as education levels, income distributions, urbanization, resource availability, and expertise. There is a need to encourage research on linkages involving factors about which less is known. It is important to define the core elements of scenarios and add other elements later.

- A large number of factors could be included in scenarios. Key foci should be on the driving forces that make a large difference and on combinations of driving forces that would create results that would be difficult to cope with. Many of these relate to the Millennium Development Goals, which could serve as an initial list of forces to consider. Toth noted that, in some cases, efforts to achieve these goals would increase vulnerability (e.g., increased access to constrained drinking water supplies). Many key drivers of impact, adaptation, and vulnerability (IAV) cannot be examined well with current models. Toth noted that the IAV assessments in the Fourth Assessment of the Intergovernmental Panel on Climate Change (IPCC) were not linked to reasonable socioeconomic scenarios.

- The group raised several concerns with existing scenarios: the lack of a baseline on which to superimpose impacts, the need to incorporate interannual variability in IAV analyses, the difficulty of assessing the effects of climate extremes on socioeconomic extremes (e.g., a drought during an economic recession), the need to develop multiple-stress scenarios and consider response options for them, and the frequent focus on climate with an implicit assumption of no change in other important drivers of well-being.

- The group discussed the question of the function of scenarios, considering that there are many stakeholders who would like analyses that are relevant to their different goals and policy concerns. It identified a need to work backward from impacts, for example, from future impacts that would be considered acceptable.
- Members questioned whether the community is organized to do all of what is needed, suggesting that an expert workshop or other tools could be considered to address this issue.

SCENARIOS FOR MITIGATION TO 2100

Michael Mastrandrea, Rapporteur

Michael Mastrandrea identified the following issues that were raised in the discussion:

- The most important factors to include in scenarios include the Kaya identity elements (population, income, energy intensity, and emissions intensity), technology availability and accessibility, governance structure (including capacity and barriers), and policy targets. Analyses of technology need to address both technical feasibility and willingness to use technologies, for example, in view of their risks. Other factors to address are changes in urban structure and patterns of consumerism that affect the carbon impact of income.
- The discussion noted several links of mitigation and adaptation with each other and with other issues (e.g., land use change, food security, water resources). For example, information needs and decisions often occur at different scales for adaptation and for mitigation. Regional issues in adaptation and mitigation priorities interact, so that analyses of only one of these get different answers from what would come from addressing both in an integrated way. A world pursuing both adaptation and mitigation can run into constraints. For example, adaptation can drive energy demand (e.g., for space cooling in hot climates, for moving water in droughts) and thus interfere with mitigation efforts. Both water and land use constraints need to be embedded in scenarios that integrate adaptation, mitigation, vulnerability, and impacts.

- Major challenges in developing scenarios include (a) the need to make global models provide local information, which raises issues about downscaling and nesting models; (b) the need to include income heterogeneity and address equity issues; (c) framing uncertainties (e.g., about technological change, consumer preferences, capacity for and implementation of governance of technology and policy, co-benefits and trade-offs); and (d) the need to consider surprises and disruptive events that may affect capacity to mitigate or to adapt.

Other issues were also raised in the discussion. One participant commented on the need to consider agriculture in mitigation analysis, saying that it is much cheaper to mitigate with changes in agricultural practices than with carbon capture and storage. Another commented that very little is known about the ultimate effects of carbon capture and storage. One participant noted that changing dietary preferences might have significant impacts on land use and greenhouse gas emissions. Another observed that macroeconomic data might lead to misunderstandings because of failure to include income gaps and generational changes in lifestyles, both of which are very important in China. Another comment concerned the different users of scenarios: different decision makers have different needs concerning adaptation and mitigation, and different information is needed in different levels and sectors. Moreover, in a society dependent on local resources, information may have to be downscaled to local ecological regions, which is different from other kinds of downscaling.

POSSIBLE PRODUCTS FOR THE FIFTH ASSESSMENT REPORT AND IMPLICATIONS FOR WORKING GROUPS 2 AND 3

Elmar Kriegler, Rapporteur

Based on the breakout group discussion, Elmar Kriegler suggested that the Fifth Assessment Report could include different levels of radiative forcing (e.g., in 2100) and the outcomes of the climate models associated with these levels, including air pollution outcomes. The report could also include representative socioeconomic pathways (RSPs) with socioeconomic data. For a given RSP and a given forcing target, it would be possible to estimate costs for mitigation and adaptation. Additional information from WG2 and WG3 would be needed to address policy options. Kriegler presented a matrix that could be filled by research. RSPs would represent adaptive capacity and mitigative capacity on two axes. Some socioeconomic variables, such as gross domestic product, affect both; some may affect only one (for example, income distribution may affect only adaptation). The effort to develop RSPs might begin with a very few variables (e.g., population and income), adding more variables after discussion (urbanization, income distribution, etc.). John Weyant said it would not be appropriate to compare cost estimates, but that it would be useful if one could compare marginal physical impacts with marginal mitigation costs (though not in a cost-benefit mode, in which physical impacts would be monetized). Steven Rose suggested that the term "RSP" may be confusing to people who might think they are representative of the range of what is possible. He suggested the term CSP, for "common socioeconomic pathways," which would indicate only that these pathways are used in common by many analyses.

WHAT THE IAV AND IAM COMMUNITIES MIGHT GET FROM EACH OTHER

Linda Mearns, Rapporteur

Prior to this discussion, Thomas Wilbanks had proposed a different frame. He said that the IAV community is not looking to the integrated assessment models (IAM) community to provide socioeconomic scenarios but would prefer to develop the scenarios together, because the communities may have different ideas about what scenarios should look like.

Linda Mearns reported that, in the breakout discussion, some participants expressed the desire for IAMs to provide regional economic insights to establish differential adaptive capacities, input on willingness to maintain adaptation infrastructures, explorations of how land use change constrains socioeconomic change, estimates of the effects of different energy mixes on the multiple stressors for a region, analyses of national security issues, and insights on "trigger points" (e.g., when do food prices get high enough to cause enough hunger to lead to political destabilization?). It was suggested that different IAMs might be able to provide information to different sectors.

Mearns reported that the discussion also identified some research opportunities: coupling demography into IAMs, analyzing two-way interactions involving land use and land cover change, and determining allowable emissions for protecting oceans from acidity.

8

Concluding Comments

Kristie Ebi offered some comments from the perspective of Working Group 2 (WG2) of the Intergovernmental Panel on Climate Change (IPCC). She said that the meeting catalyzed excellent discussions on how the two groups can work together, and she expressed the commitment of WG2 to collaborate with Working Group 3 (WG3) and Working Group 1 (WG1). She noted that many ideas had been raised at the workshop for collaboration and research, which agencies might support. However, she did not hear a proposal for a specific process for working together fast enough to get some of the research put into the IPCC Fifth Assessment. After the workshop report is published, she suggested, the next step could be a paper to propose such a process. She emphasized the need to get the scenario process started quickly, so the results can be used in the IPCC Fifth Assessment.

Timm Zwickel offered some comments from a WG3 perspective. He said that the workshop had been very enlightening. He reminded the participants of a planned meeting on story lines in October 2010 and expressed interest in an agreement on a platform for continuing this process, possibly by phone conferences, and for including the integrated assessment modeling community. Advancing on a common approach to socioeconomic scenarios might be a useful concrete task, and it might be advanced in a paper written by the community or by the leaders of WG2 and WG3.

Anthony Janetos observed that a productive avenue for national scale assessments would be to consider which factors and processes are tightly

47

or loosely coupled and suggested that Kathy Jacobs could take that issue up with U.S. government agencies. Hugh Pitcher commented that it is important to iterate, including checking ideas with the professionals in the community at least once, before publishing numbers that other people will use for analysis. He noted, for example, that there are now three sets of numbers in use for translating purchasing power parity. Brian O'Neill said that before the fall meeting on story lines, there is a need for a process to get usable input (e.g., quantified drivers) to discuss; the community needs to institutionalize the process. John Weyant mentioned thoughts about continuing the discussions this summer, suggesting that the workshop participants could hold some conference calls between now and then.

Thomas Wilbanks commented on next steps in advancing socioeconomic projections. He intended both to note things he heard and to be a catalyst to others to make suggestions. He asked participants to share ideas about next steps—both what to do and how. Some of the next steps might improve the science for the long term through tool and knowledge development. An example would be developing a library of socioeconomic scenarios, with targeted gap-filling efforts that would promote communication across the communities. He said the participants could also try to strengthen the links between climate change science and socioeconomic projection science (demography, economics, regional planning, political and institutions science, and land use modeling). They could also seek ways to incorporate technical and socioeconomic surprises.

In the near term, Wilbanks suggested the value of developing a small set of socioeconomic pathways—simple narrative story lines that can be placed aside climate change scenarios in order to assess impacts of the projected changes. More could be done to improve understanding of the coupling between climate futures and socioeconomic futures. In the next 2 years, the community could try to develop sets of scenarios to fill the space in a two-dimensional graph with high versus low climate change and high versus low socioeconomic change on its axes. It is important to discuss how the IPCC Fifth Assessment will cope with the likely proliferation of possible story lines. Finally, a shared concern with vulnerability might help the two communities communicate. Wilbanks suggested that the communities can come together most effectively by focusing on tangible issues. He ended by inviting other thoughts from the participants.

Richard Moss concluded the discussion by noting that the community does not have answers to many of the main questions people are raising. He emphasized that conclusions that cannot be supported should not be forced, noting the need to distinguish between research needs and decisions for organizing the next IPCC assessment, which must be taken by the IPCC. He concluded by thanking all participants on behalf of the organizing committee.

References

Cambridge Systematics
   2009   *Moving Cooler: An Analysis of Transportation Strategies for Reducing Greenhouse Gas Emissions.* Washington, DC: Urban Land Institute.

Coates, J.F., J.B. Mahaffie, and A. Hines
   1997   *2025: Scenarios of U.S. and Global Society Reshaped by Science and Technology.* Greensboro, NC: Oakhill Press.

Costanza, R., R. d'Arge, R. de Groot, S. Farberk, M. Grasso, B. Hannon, K. Limburg, S. Naeem, R.V. O'Neill, J. Paruelo, R.G. Raskin, P. Sutton, and M. van den Belt
   1997   The value of the world's ecosystem services and natural capital. *Nature, 387,* 253-260.

The Energy and Resources Institute
   2009   *Impacts of Relatively Severe Climate Change in North-India.* New Delhi, India: The Energy and Resources Institute.

Intergovernmental Panel on Climate Change
   2001   *Special Report on Emissions Scenarios.* Available: http://www.grida.no/publications/other/ipcc_sr/?src=/climate/ipcc/emission/ [November 2010].

Morgan, M.G., R. Cantor, W.C. Clark, A. Fisher, H.D. Jacoby, A.C. Janetos, A.P. Kinzig, J. Melillo, R.B. Street, and T.J. Wilbanks
   2005   Learning from the U.S. National Assessment of Climate Change Impacts. *Environmental Science & Technology, 39*(23), 9023-9032.

Moss, R., M. Babiker, S. Brinkman, E. Calvo, T. Carter, J. Edmonds, I. Elgizouli, S. Emori, L. Erda, K. Hibbard, R. Jones, M. Kainuma, J. Kelleher, J.F. Lamarque, M. Manning, M. Matthews, G. Meehl, L. Meyer, J. Mitchell, N. Nakicenovic, B. O'Neill, R. Pichs, K. Riahi, S. Rose, P. Runci, R. Stouffer, D. van Vuuren, J. Weyant, T. Wilbanks, J.P. van Ypersele, and M. Zurek
   2008   *Towards New Scenarios for Analysis of Emissions, Climate Change, Impacts, and Response Strategies.* Geneva: Intergovernmental Panel on Climate Change.

Moss, R.H., J.A. Edmonds, K. Hibbard, M. Manning, S.K. Rose, D.P. van Vuuren, T.R. Carter, S. Emori, M. Kainuma, T. Kram, G. Meehl, J. Mitchell, N. Nakicenovic, K. Riahi, S.J. Smith, R.J. Stouffer, A. Thomson, J. Weyant, and T. Wilbanks
  2010   The next generation of scenarios for climate change research and assessment. *Nature, 463,* 747-756.

National Research Council
  1999   *Our Common Journey: A Transition Toward Sustainability.* Board on Sustainable Development. Washington, DC: National Academy Press.
  2009a   *Informing Decisions in a Changing Climate.* Panel on Strategies and Methods for Climate-Related Decision Support. Committee on the Human Dimensions of Global Change, Division of Behavioral and Social Sciences and Education. Washington, DC: The National Academies Press.
  2009b   *New Directions in Climate Change Vulnerability, Impacts, and Adaptation Assessment: Summary of a Workshop.* J.F. Brewer, Rapporteur. Subcommittee for a Workshop on New Directions in Vulnerability, Impacts, and Adaptation Assessment. Committee on the Human Dimensions of Global Change, Division of Behavioral and Social Sciences and Education. Washington, DC: The National Academies Press.

Nelson, G.C.
  2010   Are Biofuels the Best Use of Sunlight? Pp. 15-25 in M. Khanna, J. Scheffran, and D. Zilberman, Eds., *Handbook of Bioenergy Economics and Policy.* New York: Springer.

Appendix A

Workshop Agenda and
List of Participants

Thursday, February 4, 2010

Session 1: **New Scenarios for Climate Change Research and Assessment,** *Thomas J. Wilbanks, Chair*

8:30 a.m. Workshop Objectives, Concepts, and Definitions, *Richard Moss*

8:50 Advancing the State of Science for Projecting Socioeconomic Futures, *Thomas J. Wilbanks*

9:10 Perspectives on Needs for Socioeconomic Scenarios
- Impacts, Adaptation, Vulnerability, and IPCC WG II, *Chris Field*
- Mitigation and IPCC WG III, *Ottmar Edenhofer*
- Ecosystem Services, *Anthony Janetos*
- Energy Trends and the Global Energy Assessment, *Nebojsa Nakićenović*

9:45 Relevance of the New Scenario Process, *Richard Moss*

10:15 Discussion

10:45   Coffee break

Session 2: Evolving Methods and Approaches, *Thomas J. Wilbanks,*
Chair

11:15   Philosophies and State of Science in Projecting
               Long-Term Socioeconomic Change, *Robert Lempert*
11:45   Panel Discussion: Issues in Projecting Socioeconomic
               Change
               • Demographic Change, *Thomas Buettner*
               • Economic Development, *Gary Yohe*
               • Connecting Narrative Story Lines with Quantitative
                 Socioeconomic Projections, *Ritu Mathur*
               • Quantitative Downscaling Approaches, *Tom Kram*
               • U.S. Department of Interior Scenarios

12:30 p.m.   Lunch

Session 3:   Driving Forces and Critical Uncertainties—Adaptation/
**   Vulnerability and Mitigation,** *Chris Field, Chair*

     This session will include both plenary and breakout groups that seek
to stimulate discussion of the major forces that will influence future vulnerability, adaptation potential, and mitigation potential to be analyzed in
future scenarios. Breakout groups will meet for several hours today and
reconvene over lunch on Friday.

1:30   Importance of "Driving Forces" and Critical Uncertainties
               in Scenario Construction, *M. Granger Morgan*

1:45   Panel and Open Discussion: Illustrative Drivers and
               Uncertainties for Adaptation/Vulnerability and
               Mitigation

     This session will include short (5-minute) interventions on driving
forces and disciplinary perspectives in a number of domains relevant to
assessing future vulnerability, adaptation, and mitigation. Open discussion involving all participants will follow.

               • Population, *Brian O'Neill*
               • Economy and Infrastructure, *Gary Yohe*
               • Technology, *Nebojsa Nakićenović*

- Transportation, Including Regional Planning, *Michael Replogle*
- Policy and Institutions, *Frans Berkhout*
- First and Second Best Policies, *Elmar Kriegler*
- Ecosystems and Water Resources, *Habiba Gitay*
- Food, Nutrition, and Bioenergy, *Gerald Nelson*
- Health, *Kristie Ebi*

3:45   Introduction of Breakout Groups

4:00   Break

Breakout Groups: Driving Forces and Critical Uncertainties for IAV and Mitigation

Terms of reference for breakout groups: Breakout groups are an opportunity for exchange of views on topics of interest to each group. However, each group should find time to discuss three broad sets of issues and to prepare notes and an oral report on your discussions on the following questions:

1. What factors are most important to include in socioeconomic and environmental scenarios in order to assess adaptation/mitigation?
2. How are adaptation and mitigation linked with one another and with other issues such as land use change, food security, water resources, and security, and how these linkages should be addressed in scenarios?
3. What are the major challenges in developing socioeconomic scenarios, (e.g., relating global and local/regional scales, framing uncertainties, working to an appropriate level of detail)?

4:15   Breakout Groups
   Group A: IAV 2020-2050, Chair, *Kristie Ebi*; Rapporteur,
      Linda Mearns
   Group B: IAV to 2100, Chair, *Gary Yohe*; Rapporteur,
      Ferenc Toth
   Group C: Mitigation 2020-2050, Chair, *Mikiko Kainuma*;
      Rapporteur, *Michael Mastrandrea*
   Group D: Mitigation to 2100, Chair, *Tom Kram*;
      Rapporteur, *Volker Krey*

Friday, February 5, 2010

Session 4:   **Representative Concentration Pathways (RCPs) and Socioeconomic Scenarios and Narratives,** *John Weyant, Chair*

8:30 a.m.   Characteristics, Uses, and Limits of the RCPs, *Jae Edmonds*

9:00   Multi-Model Analysis of Key Assumptions Underlying the RCPs, *Tom Kram*

9:30   Discussion

10:10   Coffee break

Session 5: Lessons from Experience, *Anthony Janetos, Chair*

10:30   Panel Discussion: Lessons from Prior and Ongoing Activities

Speakers will have approximately 10 minutes each to reflect on lessons from development and application of socioeconomic scenarios in prior assessments or planning exercises
- SRES, *Nebojsa Nakićenović*
- Millennium Ecosystem Assessment, *Gerald Nelson*
- U.S. National Assessment, *M. Granger Morgan*
- UKCIP, *Frans Berkhout*
- Private Sector, *Peter Schwartz*
- Asia Low Carbon Society Project, *Mikiko Kainuma*

11:40   Discussion

Session 6: Toward a Research Strategy, *Ottmar Edenhofer, Chair*

2:00 p.m.   Breakout groups report on driving forces and key uncertainties to be addressed in scenarios/narratives.

3:00   Panel Discussion: Developing Socioeconomic Scenarios/Narratives for Future Research and Assessment

Four speakers will give their observations on characteristics of the scenarios/narratives that need to be developed to support future research and assessments including, but not limited to, socioeconomic narratives and scenarios to complement the RCPs.

1. IAV, *Chris Field*
2. Mitigation, *Ottmar Edenhofer*
3. National adaptation assessments and planning, *Anthony Janetos*
4. International organizations, *Ian Noble*

4:00   Discussion

4:30   Next Steps in Advancing Socioeconomic Projections, *Thomas J. Wilbanks*

5:00   Adjourn

PARTICIPANTS

Ines Azevedo, Carnegie Mellon University

Martha Macedo de Lima Barata, Oswaldo Cruz Institute, Fiocruz and Centre for Integrated Studies on Environment and Climate Change, Brazil

Frans Berkhout, Institute for Environmental Studies, Free University of Amsterdam

Thomas Buettner, Population Studies Branch, U.N. Population Division/DESA, New York

Kristie Ebi, IPCC Working Group II, Technical Support, Carnegie Institution, Stanford, CA

Ottmar Edenhofer, IPCC-Working Group III, Potsdam, Germany

Jae Edmonds, Pacific Northwest National Laboratory and University of Maryland

Chris Field, Department of Global Ecology, Stanford University and Carnegie Institution, Stanford, CA

Sarah Gillig, Communication Partnership for Science and the Sea (COMPASS), Silver Spring, MD

Habiba Gitay, The World Bank, Washington, DC

Patrick Gonzalez, Center for Forestry, University of California, Berkeley

Stephane Hallegatte, CIRED/Meteo-France, Nogent-sur-Marne, France

Kathy Hibbard, Pacific Northwest National Laboratory, Richland, WA

Yasuaki Hijioka, Social and Enviromental Systems Division, National Institute for Environmental Studies, Tsukuba, Ibaraki, Japan

George Hurtt, Complex Systems Research Center, University of New Hampshire
Anthony Janetos, Joint Global Change Research Institute, Pacific Northwest National Laboratory and University of Maryland
Zou Ji, World Resources Institute, Renmin University of China, Beijing
Kejun Jiang, Energy Research Institute, China
Mikiko Kainuma, National Institute for Environmental Studies, Onogawa, Tsukuba, Japan
Robert Kopp, Office of Climate Change Policy and Technology, U.S. Department of Energy, Washington, DC
Tom Kram, Netherlands Environmental Assessment Agency, Bilthoven
Volker Krey, International Institute for Applied Systems Analysis, Laxenburg, Austria
Elmar Kriegler, Potsdam Institute for Climate Impact Research, Potsdam, Germany
Hadas Kushnir, The National Academies, Washington, DC
Robert Lempert, Rand Corporation, Santa Monica, CA
Marc Levy, Center for International Earth Sciences Information Network, Columbia University
Michael MacCracken, Climate Institute, Washington, DC
Michael Mastrandrea, Carnegie Institution for Science, Stanford University
Ritu Mathur, The Energy and Resources Institute, New Delhi, India
Patrick Matschoss, IPCC-TSU-Working Group III, Potsdam, Germany
Linda Mearns, National Center for Atmospheric Research, Boulder, CO
Nobou Mimura, Center for Water Environment Studies Ibaraki University, Hitachi, Ibaraki, Japan
M. Granger Morgan, Carnegie Mellon University
Richard Moss, Joint Global Change Research Institute, University of Maryland
Nebojsa Nakićenović, International Institute for Applied Systems Analysis, Laxenburg, Austria
Gerald Nelson, International Food Policy Research Institute, Washington, DC
Ian Noble, The World Bank, Washington, DC
Robert O'Connor, National Science Foundation, Arlington, VA
Cara O'Donnell, Science and Technology Policy Institute, Washington DC
Brian O'Neill, National Center for Atmospheric Research, Boulder, CO
Jon Padgham, START, Washington, DC
Michael Replogle, Institute for Transportation and Development Policy, Washington, DC
Steven Rose, Global Climate Change Policy Resource Center, Electric Power Research Institute, Palo Alto, CA

Steffen Schloemer, IPCC-Working Group III, Potsdam, Germany
Peter Schwartz, Global Business Network, San Francisco, CA
Avery Sen, Office of Program Planning and Integration, National
  Oceanographic and Atmospheric Administration, Silver Spring, MD
P.R. Shukla, Indian Institute of Management, Vastrapur, Ahmedabad,
  India
Paul C. Stern, The National Academies, Washington, DC
Miron Straf, The National Academies, Washington, DC
Massimo Tavoni, Princeton Environmental Institute, Princeton
  University
Allison Thompson, Joint Global Change Research Institute, College
  Park, MD
Ferenc Toth, Department of Nuclear Energy, International Atomic
  Energy Agency, Vienna, Austria
Robert Vallario, U.S. Department of Energy, Washington, DC
Detlef van Vuuren, Netherlands Environmental Assessment Agency,
  Bilthoven
Hassan Virji, START, Washington, DC
Thanh Vo Dinh, Office of Program Planning and Integration, National
  Oceanographic and Atmospheric Administration, Silver Spring, MD
Brian Wee, NEON, Inc. Boulder, CO
Leigh Welling, U.S. National Park Service, Washington, DC
John Weyant, Department of Management Science and Engineering,
  Stanford University
Thomas J. Wilbanks, Oak Ridge National Laboratory, Oak Ridge, TN
Gary Yohe, Department of Economics, Wesleyan University
Timm Zwickel, IPCC-TSU-Working Group III, Potsdam, Germany

Appendix B

Biographical Sketches of Panel Members and Staff

RICHARD H. MOSS is a senior staff scientist at the Joint Global Change Research Institute. He was previously vice president and managing director for climate change at the World Wildlife Fund (WWF). His recent work includes developing conservation plans that account for changing climate and reduce greenhouse gas emissions and developing the WWF role on adapting to climate change. He has served the Intergovernmental Panel on Climate Change (IPCC) as head of the technical staff of the impacts-adaptation-mitigation working group (1993-1999); as editor or coauthor of a number of IPCC reports, including the panel's first examination of *The Regional Impacts of Climate Change* (1998); and as contributor to the 2007 Nobel prize-winning IPCC assessment. He also coauthored IPCC's first methodology on consistently evaluating and communicating scientific uncertainty in assessments, used by authors of the IPCC's Third Assessment Report. He currently serves as cochair of the IPCC Task Group on Data and Scenario Support for Impact and Climate Analysis. From 2000 to 2006, he directed the coordination office for the United States Climate Change Science Program, leading preparation of the program's 10-year *Strategic Plan* (2003), which focuses on development and application of research to support decision making. He has a Ph.D. from Princeton University's Woodrow Wilson School of Public and International Affairs, an M.P.A. from Princeton University, and a B.A. from Carleton College.

KRISTIE L. EBI is executive director of IPCC Working Group II Technical Support at the Carnegie Institution. She is an epidemiologist who has

59

worked in the field of global climate change for 10 years. Her research focuses on potential impacts of climate variability and change, including impacts associated with extreme events, thermal stress, food-borne diseases, and vector-borne diseases, and on the design of adaptation response options to reduce current and projected future negative impacts. She is chief editor of the book *Integration of Public Health with Adaptation to Climate Change: Lessons Learned and New Directions.* She was a lead author for the human health chapter of the IPCC Fourth Assessment Report; a convening lead author on the World Health Organization publication *Methods of Assessing Human Health Vulnerability and Public Health Adaptation to Climate Change*; and lead author in the Millennium Ecosystem Assessment and the U.S. National Assessment of the Potential Consequences of Climate Variability and Change. She has more than 25 years of multidisciplinary experience in environmental issues and numerous publications. Her scientific training includes an M.S. in toxicology, Ph.D. and M.P.H. degrees in epidemiology, and two years of postgraduate research in epidemiology at the London School of Hygiene and Tropical Medicine.

KATHY A. HIBBARD is senior program manager at the Pacific Northwest National Laboratory (PNNL) and executive officer for the International Geosphere-Biosphere Programme's (IGBP) Earth system project, the Analysis, Integration and Modeling of the Earth System (AIMES). At PNNL, she is leading a new initiative to develop an integrated regional climate, socioeconomic, and energy systems model. Her major area of scientific interest is understanding the consequences of disturbance (natural and anthropogenically forced) to terrestrial biogeochemical cycles through field observations and modeling. Her primary focus in the AIMES project is to understand and integrate human-environmental processes (e.g., land use, emissions) in Earth system modeling. She has authored or coauthored numerous publications and two book chapters and has worked in international program development for the IGBP's Global Carbon Project and IGBP/GAIM Task Force. She has been a member of the Ecological Society of America since 1991 and served AGU Biogeosciences from 2001 to 2004 as fall meeting program committee representative. She has B.S. and M.S. degrees from Colorado State University in biology and range science and a Ph.D. from Texas A&M University in range ecology and management.

ANTHONY C. JANETOS is director of the Joint Global Change Research Institute at the University of Maryland. He previously served as vice president of the H. John Heinz III Center for Science, Economics and the Environment, where he directed the center's Global Change Program. He has written and spoken widely to policy, business, and scientific audiences on the need for scientific input and scientific assessment in the policy-making process and about the need to understand the scientific, environmental, economic, and policy linkages among the major global environmental issues. He has served on several national and international study teams, including working as a co-chair of the U.S. National Assessment of the Potential Consequences of Climate Variability and Change. He also was an author of the IPCC's Special Report on Land-Use Change and Forestry, the Global Biodiversity Assessment, and a coordinating lead author in the recently published Millennium Ecosystem Assessment. He is a member of the National Research Council's (NRC) Climate Research Committee. Janetos graduated magna cum laude from Harvard College with a bachelor's degree in biology and has master's and Ph.D. degrees in biology from Princeton University.

MIKIKO KAINUMA is chief of the Climate Policy Assessment Research Section at the National Institute for Environmental Studies (NIES). She has been developing the Asia-Pacific Integrated Model (AIM) with Kyoto University and several other institutes across Asia, including China, India, Korea, and Thailand. She leads the Low-Carbon Asia Research Project, funded by the Global Environmental Research Fund of the Ministry of Environment of Japan. Since 1977 she has worked on air pollution and climate change at NIES. She was a lead author of the IPCC's Fourth Assessment Report. She was a member of IPCC Task Group on New Emissions Scenarios for a possible IPCC Fifth Assessment Report. In addition she has worked on United Nations Environment Programme/Global Environment Outlook scenarios. She is an adjunct professor at Japan Advanced Institute of Science and Technology. She has B.S., M.S., and Ph.D. degrees in applied mathematics and physics from Kyoto University.

RITU MATHUR is associate director of the Energy Environment Policy Division at The Energy and Resources Institute (TERI). An economist by training, she has used various modeling and analytical tools for developing national and sectoral level energy models to examine the prospects of future energy use patterns and their implications on the economy and environment. Over the past 15 years, she has led several projects with interdisciplinary teams, addressing such cross-cutting issues as energy pricing, environmental implications of energy use, examining mitigation options, and the potential for the country to combat climate change. She has authored several papers related to energy use and its implications on the environment at the local and global levels. She has been a key discussant at various international fora on topics related to developing country perspectives toward climate change, mitigation prospects for India, and energy-environment policy. She was part of the IPCC Fourth Assessment Report approval process, in which she represented the India delegation, and has also participated at various side events at the Congressional Oversight Panel and Subsidiary Body for Scientific and Technological Advice meetings on issues of interest in the international negotiation processes. She has a Ph.D. in energy science from Kyoto University.

NEBOJSA NAKIČENOVIČ is deputy director of the International Institute for Applied Systems Analysis (IIASA), professor of energy economics at the Vienna University of Technology, and director of the Global Energy Assessment. He is also a member of the United Nations Secretary General Advisory Group on Energy and Climate Change; the Advisory Council of the German Government on Global Change; the Advisory Board of the World Bank's *World Development Report 2010: Development and Climate Change*; the International Council for Science Committee on Scientific Planning and Review; and the Global Carbon Project. He was a convening lead author of the IPCC's Second Assessment Report, its Special Report on Emissions Scenarios, and the World Energy Assessment: Energy and the Challenge of Sustainability; a coordinating lead author of the IPCC's Fourth Assessment Report and of the Millennium Ecosystem Assessment; lead author of the IPCC's Third Assessment Report; director of Global Energy Perspectives at the World Energy Council; a member of the International Science Panel on Renewable Energies; and guest professor at the Technical University of Graz. Among his research interests are the long-term patterns of technological change, economic development and response to climate change, and, in particular, the evolution of energy, mobility, information, and communication technologies. He has B.S. and M.S. degrees in economics and computer science from Princeton University and the University of Vienna, where he also completed a Ph.D. He also has an honoris causa Ph.D. degree in engineering from the Russian Academy of Sciences.

PAUL C. STERN is a principal staff officer at the National Academy of Sciences (NAS)/NRC, director of its Committee on the Human Dimensions of Global Change, and study director for this panel. His research interests include the determinants of environmentally significant behavior, particularly at the individual level; participatory processes for informing environmental decision making; processes for informing environmental decisions; and the governance of environmental resources and risks. He is coauthor of the textbook *Environmental Problems and Human Behavior, Second Edition* (2002); coeditor of numerous NRC publications, including *Public Participation in Environmental Assessment and Decision Making* (2008), *Decision Making for the Environment: Social and Behavioral Science Priorities* (2005), *The Drama of the Commons* (2002), *Making Climate Forecasts Matter* (1999), *Environmentally Significant Consumption: Research Directions* (1997),

Understanding Risk (1996), *Global Environmental Change: Understanding the Human Dimensions* (1992), and *Energy Use: The Human Dimension* (1984). He directed the study that produced *Informing Decisions in a Changing Climate* (2009). He coauthored the article "The Struggle to Govern the Commons," which was published in *Science* in 2003 and won the 2005 Sustainability Science Award from the Ecological Society of America. He is a fellow of the American Association for the Advancement of Science and the American Psychological Association. He holds a B.A. from Amherst College and an M.A. and Ph.D. from Clark University, all in psychology.

THOMAS J. WILBANKS is a corporate research fellow at the Oak Ridge National Laboratory and leads its Global Change and Developing Country Programs. A past president of the Association of American Geographers, he conducts research on such issues as sustainable development, energy and environmental technology and policy, responses to global climate change, and the role of geographical scale in all of these regards. Coedited recent books include *Global Change and Local Places* (2003), *Geographical Dimensions of Terrorism* (2003), and *Bridging Scales and Knowledge Systems: Linking Global Science and Local Knowledge* (2006). Wilbanks is a member of the NRC's Board on Earth Sciences and Resources, chair of the NRC's Committee on Human Dimensions of Global Change, and a member of a number of other NAS/NRC activities. He is a coordinating lead author for the IPCC's Fourth Assessment Report, Working Group II, Chapter 7 (Industry, Settlement, and Society); coordinating lead author for the Climate Change Science Program's Synthesis and Assessment Product (SAP) 4.5 (Effects of Climate Change on Energy Production and Use in the United States); and lead author for one of three sections (Effects of Global Change on Human Settlements) of SAP 4.6 (Effects of Global Change on Human Health and Welfare and Human Systems). He has a B.A. in social sciences from Trinity University and M.A. and Ph.D. degrees in geography from Syracuse University.